# Biotechnology

## A Laboratory Course

### Second Edition

# Biotechnology _____

## A Laboratory Course
### Second Edition

**Jeffrey M. Becker**
*Department of Microbiology*
*Department of Biochemistry, Cellular,*
*  and Molecular Biology*
*The University of Tennessee*
*Knoxville, Tennessee*

**Guy A. Caldwell**
Department of Biological Sciences
Columbia University
New York, New York

**Eve Ann Zachgo**
Department of Molecular Biology
Massachusetts General Hospital
Boston, Massachusetts

**Academic Press**
San Diego   New York   Boston   London   Sydney   Tokyo   Toronto

*Front cover photograph*: Bacteria and yeast, the workhorses of biotechnology and molecular biology, are depicted as a bacterial colony and one budding yeast cell containing plasmid DNA and the DNA double helix. This illustration represents the manipulation of living organisms to produce material useful for humans — a definition of biotechnology in the broadest sense. The recent revolution in this field of endeavor has been made possible by the explosion of knowledge in molecular biology and by the rapid proliferation of techniques using microorganisms and recombinant DNA.

Academic Press
A Division of Harcourt Brace & Company
525 B Street, Suite 1900, San Diego, California 92101-4495

*United Kingdom Edition published by*
Academic Press Limited
24-28 Oval Road, London NW1 7DX

Library of Congress Cataloging-in-Publication Data

Becker, Jeffrey M.
    Biotechnology : a laboratory course / by Jeffrey M. Becker, Guy A.
Caldwell, Eve Ann Zachgo, -- 2nd ed.
        p.        cm.
    Includes bibliographical references and index.
    ISBN 0-12-084562-8 (alk. paper)
    1. Biotechnology--Laboratory manuals.    I. Caldwell, Guy A.
II. Zachgo, Eve Ann.    III. Title.
TP248.2.B374    1996
660'.6'078--dc20                                             95-41427
                                                             CIP

PRINTED IN THE UNITED STATES OF AMERICA
        98  99   00   01   EB   9   8   7   6   5   4   3

To the wife of Jeff Becker,
   Nancy Becker
and to his children,
   Rachel, Benjamin, and Sarah Becker
       and
   Jeffrey, Daniel, and Deborah Cohen

To the wife of Guy Caldwell,
   Kim Caldwell

To the loved ones of Eve Zachgo,
   They know who they are

# Contents _____

Preface to the Second Edition    xiii
Preface to the First Edition    xv
Acknowledgments    xvii
Suggested Schedule for Exercises    xix

## Introductory Notes
### Record Keeping and Safety Rules

Format of Student Laboratory Records    3
The Ten Commandments of Record Keeping    5
Safety Rules in the Laboratory    7

## Exercise 1
### Aseptic Technique and Establishing Pure Cultures: The Streak Plate and Culture Transfer    9

## Exercise 2
### Preparation of Culture Media    17

## Exercise 3
### The Growth Curve    25

## Exercise 4
### Isolation of Plasmid DNA from *Escherichia coli*: The Mini-Prep    31

*Exercise 5*

     Purification, Concentration, and Quantitation of DNA     37

*Exercise 6*

     Large-Scale Isolation of Plasmid DNA by
     Column Chromatography     49

*Exercise 7*

     Amplification of a *lacZ* Gene Fragment by the Polymerase
     Chain Reaction     55

*Exercise 8*

     Restriction Digestion and Agarose Gel Electrophoresis     63

*Exercise 9*

     Southern Transfer     75

*Exercise 10*

     Preparation, Purification, and Hybridization of Probe     85

*Exercise 11*

     Transformation of *Saccharomyces cerevisiae*     97

*Exercise 12*

     Isolation of Plasmid from Yeast and *Escherichia
coli* Transformation     105

*Exercise 13*

     Protein Assays     119

**Exercise 14**

Qualitative Assay for $\beta$-Galactosidase in
Yeast Colonies    125

**Exercise 15**

Determination of $\beta$-Galactosidase in Permeabilized
Yeast Cells    131

**Exercise 16**

Assay of $\beta$-Galactosidase in Cell Extracts    135

**Exercise 17**

$\beta$-Galactosidase Purification    141

**Exercise 18**

Western Blot: Probe of Protein Blot with Antibody
to $\beta$-Galactosidase    157

**Appendix 1**

Alternative Protocols and Experiments

**Exercise  1A**  Isolation and Characterization of Auxotrophic
Yeast Mutants    165
**Exercise  2A**  Measurement of pH    170
**Exercise  3A**  Use of the Spectrophotometer    174
**Exercise  6A**  Isolation of Plasmid DNA: The Maxi-Prep    178
**Exercise 10A**  Colony Hybridization    189

**Appendix 2**

Buffer Solutions    192

**Appendix 3**
Preparation of Buffers and Solutions    195

**Appendix 4**
Properties of Some Common Concentrated Acids
and Bases    200

**Appendix 5**
Use of Micropipettors    201

**Appendix 6**
Safe Handling of Microorganisms    204

**Appendix 7**
List of Cultures    207

**Appendix 8**
Storage of Cultures and DNA    208

**Appendix 9**
Sterilization Methods    210

**Appendix 10**
Preparation of Stock Solutions for Culture Media    212

**Appendix 11**
Growth in Liquid Medium    214

**Appendix 12**
Determination of Viable Cells    217

**Appendix 13**

    Determination of Cell Mass    219

**Appendix 14**

    Determination of Cell Number    221

**Appendix 15**

    Nomenclature of Strains    223

**Appendix 16**

    Glassware and Plasticware    226

**Appendix 17**

    Preparation of Tris and EDTA    228

**Appendix 18**

    Basic Rules for Handling Enzymes    231

**Appendix 19**

    Effects of Common Contaminants on Protein Assays    235

**Appendix 20**

    Manufacturers' and Distributors' Addresses    237

**Appendix 21**

    Surfing the Bionet: World Wide Web Addresses    245

    Glossary    247

    Index    253

# *Preface to the Second Edition*

Since publication of the first edition in 1990, *Biotechnology: A Laboratory Course* has been used for laboratory courses at undergraduate and graduate levels in many colleges and universities in the United States and has been translated into Chinese.

The objectives of this manual are unchanged. We have attempted to create a text that consists of a series of laboratory exercises providing a continuum of experiments. We begin with basic techniques and culminate in the utilization of previously acquired technical experience and experimental material. Two organisms (*Saccharomyces cerevisiae* and *Escherichia coli*) a single plasmid (pRY121) and a single enzyme (β-galactosidase) are the experimental material, yet the procedures and principles demonstrated are widely applicable to other systems.

The second edition differs from the first in many ways. Exercises 1 through 6 have been consolidated and streamlined. New procedures are used for large-scale plasmid isolation, yeast transformation, and DNA quantitation. New exercises have been introduced for polymerase chain reaction (PCR), for detecting β-galactosidase in yeast colonies, for shuttling plasmids from *S. cerevisiae* to *E. coli,* and for detecting proteins on blots by antibodies. The Glossary has been amended and several new Appendixes have been added. For example, a new Appendix contains a list of World Wide Web addresses enabling students and instructors to access valuable information on the biological sciences on the Internet. All of the remaining Exercises and Appendixes have been completely revised to incorporate suggestions from students and instructors, and the references at the end of each exercise have been expanded, updated, and annotated.

We trust that this new edition will continue to serve as an aid to the establishment or instruction of laboratory courses in biotechnology and molecular biology.

Jeffrey M. Becker
Guy A. Caldwell
Eve Ann Zachgo

# *Preface to the First Edition*

This book is different. To our knowledge, there is no text designed for instruction at the introductory or even advanced level that is comparable to this manual. We have attempted to create a text that consists of a series of laboratory exercises, beginning with basic techniques and culminating in the utilization of previously acquired technical experience and experimental material. Thus, this manual provides a continuum of experiments using the same biological material throughout, which is beneficial in understanding a common sequence of technical stages used in biotechnology in a framework which is suited to instructional methodologies.

This manual is an outgrowth of an introductory laboratory course for senior undergraduate and first year graduate students in the biological sciences at The University of Tennessee, which was designed to provide students with an in-depth experience and understanding of selected methods, techniques, and instrumentation used in modern biotechnology. A continuum was established by using two principal organisms: the bacterium *Escherichia coli* and the yeast *Saccharomyces cerevisiae*. In addition, a single plasmid (pRY121) and a single enzyme (β-galactosidase) for demonstrating various biotechnological manipulations were utilized. Because our main objective was to provide a step-by-step laboratory course for students, there are aspects of molecular biology and biotechnology which are not included.

To facilitate the use of this manual by diverse undergraduate institutions, we have provided a suggested weekly schedule based on a fifteen-week format, which we have found useful at The University of Tennessee; ensured ready availability of the few microbial cultures (*Escherichia coli* and *Saccharomyces cerevisiae*) necessary for this course; attempted to design the course for a minimum number of reagents and supplies; and do not propose work with radioactivity.

The continued impact of recombinant DNA technology in both academic and industrial research has resulted in a substantial change in the curriculum of graduate and undergraduate biology programs. This increased emphasis on new technology required a novel and contemporary method of instruction. It is our hope that this manual will aid academic institutions in their efforts to establish a laboratory course in molecular biology and biotechnology.

Jeffrey M. Becker
Guy A. Caldwell
Eve Ann Zachgo

# *Acknowledgments*

We would like to acknowledge the following individuals and organizations for their time, effort, and support in the preparation of the second edition of this manual: Frances Cox for her dedicated service in typing and editing; Paula Keaton for typing the first edition and assisting in the revision of the second edition; Greg Abel, Kumar Alagramam, Kim Caldwell, Angus Dawe, Angela McKinney, Mohan Solway, and Tim Wiltshire for their useful advice and suggestions on the experimental protocols; and the Science Alliance, State of Tennessee Center for Excellence in the Life Sciences, for the laboratory facilities and funding of the biotechnology laboratory at the University of Tennessee, Knoxville. We are grateful for the continuous support of Ms. Shirley Light of Academic Press, whose encouragement and interest allowed us to undertake this revision.

# Suggested Schedule for Exercises _____

## Instructor's Note _____

The schedule below is intended to help you plan the course. Alternative exercises are provided as supplementary material.

| Week | Day 1 | Day 2 | Day 3 | Day 4 |
|------|-------|-------|-------|-------|
| 1 | Introduction/ Orientation[†] | | Exercise 1* Exercise 2 | |
| 2 | Exercise 3, Day 1 | | Exercise 3, Day 3 Exercise 4, Day 3 | |
| 3 | Exercise 5 | | Exercise 6** | |
| 4 | Exercise 7 | | Exercise 8 | |
| 5 | Exercise 9, Day 1 | Exercise 9, Day 2 | Exercise 9, Day 3 | |
| 6 | Exercise 10, Day 1 | | Exercise 10, Day 2 | |
| 7 | Exercise 10,[††] Day 3 | Exercise 10, Day 4 | Exercise 10, Day 5 | |

* Exercise 1A (Isolation and Characterization of Auxotrophic Yeast Mutants) is an additional exercise. (See Appendix 1.)

[†] Depending on the experience of the students, Exercise 2A (Measurement of pH) and/or 3A (Use of the Spectrophotometer) may be added to Day 1. (See Appendix 1.)

** Allow 2 days for Exercise 6A (Isolation of Plasmid DNA: The Maxi-Prep). (See Appendix 1.)

[††] Allow 7 days for Exercise 10A (Colony Hybridization). (See Appendix 1.)

| | | | |
|---|---|---|---|
| 8 | Exercise 11 | | Exercise 12, Day 1 | |
| 9 | Exercise 12, Day 3 | | Exercise 12, Day 4 | |
| 10 | Exercise 13 | Exercise 14, Day 1 | Exercise 14, Day 2 | Exercise 14, Day 3 |
| 11 | Exercise 15 | | Exercise 16 | |
| 12 | Exercise 17A, Day 1 | | Exercise 17A, Day 2 | |
| 13 | Exercise 17B, Day 1 | Exercise 17B, Day 2 | Exercise 17B, Day 3 | |
| 14 | Exercise 17C, Day 1 | Exercise 17C, Day 2 | Exercise 18 | |

# Record Keeping and Safety Rules

Format of Student Laboratory Records

The Ten Commandments of Record Keeping

Safety Rules in the Laboratory

# Format of Student Laboratory Records

These guidelines should be followed to record experiments correctly.

## A. Purpose/Introduction

In two or three sentences state exactly the purpose(s) and objective(s) of the experiment; what was done, to what, and why. Do not include as a part of this abstract familiarization with equipment or techniques, unless applicable. A *short* conclusion stating results/ideas may be added, if applicable. THIS IS NOT TO BE LENGTHY!

## B. Materials and Methods

Include here *only* any deviations, or additional equipment/procedures, from those stated in the laboratory manual. Otherwise, merely write, "No deviations from the planned procedure were performed." However, accurate and complete cross-referencing to any references or procedures is mandatory.

## C. Data/Results

*All* data generated should be recorded *during* the experiment. This includes recopying any tables, graphs, formulas, etc., from the laboratory manual and/or making new tables, graphs, as necessary, to represent the data accurately and *neatly*. Also, this section should include all calculations, averages, error analyses, and corrections of recorded data.

## D. Discussion and Conclusions

This section should include interpretations, conclusions, or suggestions regarding the results obtained. If applicable, include the expected results, and discuss why they were or were not achieved. State evidence for your views, including any assumptions you have made. THIS IS *NOT* TO BE A SUMMARY OF THE ENTIRE EXPERIMENT— ONLY A DISCUSSION OF THE FINAL RESULTS!

## E. References

Include any that were consulted for the experiment or made in reference to the report. This is usually *at least* your laboratory manual.

### Some Important Notes _____

1. *At all times* be honest and concise.

2. Label *all* tables and graphs to indicate what they represent.

3. DO NOT use different color pens or pencils in any portion of the reports.

4. Graphs should be numbered on axes in divisions that are easy to work with (i.e., 5, 10, 15, 20, etc.; not 2.3, 4.6, 6.9, etc.).

5. *Remember,* this is *not* a personal diary and should not include any personal notations.

6. *Remember,* you are *not* being graded on your results; therefore, an accurate representation of them, with an intelligent explanation or hypothesis for any data obtained, is far more important.

7. If available, word processing and spreadsheet software will facilitate report writing and data analysis.

# The Ten Commandments of Record Keeping*

1. Keep the record factual. Don't editorialize.

2. Use a record book with permanent binding to avoid page deletions and insertions.

3. Make two copies of all notebook entries, one of which should be kept safely at a separate location.

4. Enter data and information directly into the record book promptly as generated. You may wish to sign and date each page of the record book at the time of entry (signing is required procedure for industrial research notebooks). Do not rely on memory or informal loose sheets for entries with the intention of later putting these into the bound record book.

5. Use permanent ink, preferably black, which will reproduce well when photocopied.

6. Identify errors and mistakes and explain them.

7. Attach support records to the record book or store such records, after properly referencing and cross-indexing, in a readily retrievable manner.

* *Derived from* "Record Books: Their Generation, Maintenance, and Safekeeping." Dow Chemical Company, Midland, Michigan.

8. Use standard accepted terms; avoid abbreviations, code names, or code numbers if possible.

9. Keep the record book clean; avoid spills and stains.

10. Keep a table of contents and index the record book as soon as it is filled.

# Safety Rules in the Laboratory

1. Prepare for each laboratory period by reading each exercise and becoming familiar with the principles and methods involved. By being familiar with the exercise you decrease the chances of an accident. Also, advance preparation allows you to use your time efficiently in the laboratory to complete the experiment.

2. No eating, drinking, or smoking is permitted in the laboratory.

3. Laboratory coats or aprons must be worn at all times in the laboratory. This is to ensure that culture material is not accidentally deposited on your clothes or skin, and as a safeguard to protect your clothes and yourself from chemical spills and stains.

4. Only those materials pertinent to your laboratory work, such as laboratory manuals, laboratory notebooks, and other laboratory materials, should be brought to your laboratory work space. All other items, such as coats, books, and bags, should be stored away from your work area.

5. Begin each laboratory session by disinfecting your work area. Saturate the area with a disinfectant, spread the disinfectant with a paper towel, and allow the area to dry. Repeat this procedure after you have finished your work to ensure that any material you have deposited on the work surface is properly disinfected.

6. All culture material and chemicals should be properly labeled with your name, class, date, and experiment. Labeling is critical to avoid improper use or disposal of material.

7. Be very careful with Bunsen burners. To avoid injuries, burners should be turned off when not in use. When reaching for objects, be careful not to place your hands into the flame.

8. All contaminated material must be disinfected before disposal or reuse. All material to be autoclaved should be placed in a proper receptacle for collection. Used pipets should be placed in disinfectant.

9. After the laboratory session, observe good hygiene by washing your hands before leaving the laboratory.

10. In the event of any accident or injury, report immediately to the laboratory instructor so that prompt and proper action can be taken.

## Instructor's Note

Students should be aware that safety issues are of the utmost importance in the laboratory. Time should be spent during the first laboratory meeting to familiarize students with the placement of eye wash-stands, showers, fire extinguishers, fire exits, and hazardous waste disposal sites.

# 1

# Aseptic Technique and Establishing Pure Cultures: The Streak Plate and Culture Transfer

## Introduction

The utilization of microbes in biotechnology depends on pure cultures, which consist of only a single species, and the maintenance of the purity of the isolates through subsequent manipulations (see Appendix 8, Storage of Cultures and DNA). Most applications in biotechnology involve the use of pure cultures.

Most methods for obtaining pure cultures rely on some form of dilution technique. The most useful and pragmatic method is the streak plate, in which a mixed culture is spread or streaked over the medium surface in such a way that individual cells become separated from one another. Each isolated cell grows into a colony and, therefore, a pure culture (or clone) because cells are the progeny of the original single cell.

There are several checks to establish the purity of a culture.

1. On restreaking, an isolated colony from an initial streak plate should yield only a single type of isolated colony whose colonial morphology is consistent with the initial isolate.

2. Microscopic examination of the organisms from a colony should reveal only a single type of cell; differential staining procedures, such as the gram stain, are useful for establishing that the colony does not contain a mixture of different microbial types.

There is more than one method for obtaining a good streak plate and each method requires some practice. It is important to remember that the more cells one starts with on the inoculating loop, the more streaking (dilution) is required. It is not necessary to start with large amounts or "gobs" of culture material on loops. For this laboratory exercise, we will use two different streak plate procedures.

Aseptic technique is required to transfer pure cultures and to maintain sterility of media and solutions (see Appendix 6, Safe Handling of Microorganisms). By aseptic technique the biotechnologist takes prudent precautions to prevent contamination of the culture or solutions by unwanted microbes. Many of the petri dishes and tissue culture plates that are used for growing pure cultures of microorganisms are made of plastic and come presterilized from the manufacturer; filling these vessels with a sterile medium requires the use of aseptic technique. Proper aseptic transfer technique also protects the biotechnologist from contamination with the culture, which should always be treated as a potential pathogen. Aseptic technique involves avoiding any contact between the pure culture, sterile medium, and sterile surfaces of the growth vessel with contaminating microorganisms. To accomplish this task, (1) the work area is cleansed with an antiseptic to reduce the numbers of potential contaminants; (2) the transfer instruments are sterilized; for example, the transfer loop is sterilized by heating with a Bunsen burner before and after transferring; and (3) the work is accomplished quickly and efficiently to minimize the time of exposure during which the contamination of the culture or laboratory worker can occur.

The typical steps for transferring a culture from one vessel to another are as follows: (1) flame the transfer loop; (2) open and flame the mouths of the culture tubes; (3) pick up some of the culture growth and transfer it to the fresh medium; (4) flame the mouths of the culture vessels and reseal them; and (5) reflame the inoculating loop. Essentially the same technique is used for inoculating petri dishes, except that the dish is not flamed, and for transferring microorganisms from a culture vessel to a microscope slide.

Developing a thorough understanding and knowledge of aseptic technique and culture transfer procedures is a prerequisite for working with microbiological cultures. You will save yourself a lot of time and energy and will avoid erroneous results if a few simple and commonsense rules are observed when working with cultures.

1. Always sterilize the inoculating loop by flaming before using it to enter any culture material.

2. Always flame the lip of the culture tube before inserting the sterile loop into the culture. This destroys any contaminating cells that may have been inadvertently deposited near the lip of the tube during previous transfer or by other means.

3. Keep all culture materials covered with their respective caps and lids when not making transfers. Do not lay tube caps or petri dish lids on the tabletop, thereby exposing cultures to possible contamination. When transferring colonies from petri plates, use the lid as a shield by slightly raising it enough so that the loop can be inserted but the agar surface is still protected from contaminants falling on it.

4. Do not allow tube closures or petri dish lids to touch anything except their respective culture containers. This will prevent contamination of closures and therefore of cultures.

5. Use proper handling procedures for closure removal and return.

## Reagents/Supplies

Inoculating loops
LB agar plates [for LB agar, add agar (15 g/liter) to LB broth]
LB broth (Luria–Bertani broth: tryptone, 10 g/liter; yeast extract, 5 g/liter; NaCl, 10 g/liter; pH adjusted dropwise to 7.5 with NaOH)

Mixed culture containing *Escherichia coli* and *Saccharomyces cerevisiae*

Pure broth cultures of *E. coli* (strain LE392) and *S. cerevisiae* (strain YNN281) (see Appendix 7 for information on how to obtain these cultures)

Test tube racks

Test tubes (18 × 150 mm)

YEPD agar plates (yeast extract–peptone–dextrose; yeast extract, 10 g/liter; peptone, 20 g/liter; dextrose, 20 g/liter; agar, 20 g/liter)

## Equipment

Bunsen burners

Incubator at 30°C

Incubator at 37°C

### Instructor's Note _____

One day prior to the laboratory exercise inoculate 5 ml of LB broth with *E. coli* and 5 ml of YEPD with *S. cerevisiae* (one culture per student). Grow *E. coli* at 37°C and *S. cerevisiae* at 30°C. On the day of the laboratory exercise, also prepare mixed cultures (one per student) by adding 1 ml of *E. coli* and 1 ml of *S. cerevisiae* from the overnight cultures to a sterile test tube.

## Procedure

### Part A. The Streak Plate

1. Label five LB and five YEPD agar plates with your name, date, class, and the name of the source culture used.

2. Use the quadrant streak method (described below) to prepare two streak plates (LB agar and YEPD agar) of the mixed culture. Repeat the procedure, using the continuous streak method (described below). Also streak (by both streaking methods) pure cul-

tures of *E. coli* onto LB agar and *S. cerevisiae* onto YEPD agar. Be sure to label each plate.

## Quadrant Streak Method (see Figure 1.1)

a.  Draw quadrants on the outside bottom of an agar petri plate.

b.  Flame an inoculating loop and be sure to allow the loop to cool before introducing it into the broth culture. (Hot loops kill microorganisms and produce aerosols when they touch cool agar.)

c.  Allow the loop to touch the surface of the agar lightly and slide it gently over the surface in one quadrant in a continuous streaking motion. Use the petri dish cover to protect the agar surface and prevent contamination from falling onto the medium. Avoid digging the loop into the agar. Use reflected light to see where you have streaked and inoculated your plate. These areas will appear as faint scratch marks. This will allow you to position your inoculations better.

d.  Flame the loop, allow it to cool, and cross-streak from the previous area to the second quadrant. Always sterilize the loop after inoculating each section of the plate; this will kill any cells adhering to the loop and prevent contamination of the next inoculation.

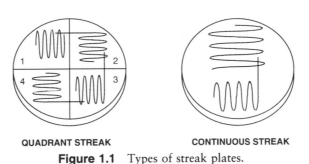

QUADRANT STREAK                    CONTINUOUS STREAK

**Figure 1.1**  Types of streak plates.

   e. Repeat this same procedure for each succeeding new area, until all four quadrants on the plate are inoculated.

   f. Flame the loop when you have finished.

## Continuous Streak Method (see Figure 1.1)

   a. Take a small amount of culture inoculum on the loop and spread it in a single, continuous, back-and-forth motion over one-half of the plate.

   b. Without flaming or lifting the loop and using the same face of the inoculating loop, turn the plate 180° and continue the streaking procedure as done in the initial area.

3. Prepare a control or sham inoculation in which you use only the sterilized inoculating loop without culture to prepare a streak plate by either streaking procedure. Such a plate will be a good indicator of your aseptic technique, as nothing should grow on this plate.

4. Invert the plates to prevent water condensate from spreading bacteria over the agar surface, place your plates in a 37°C incubator for *E. coli* and LB mixed culture and in a 30°C incubator for *S. cerevisiae* and YEPD mixed culture, and incubate for 24–48 hours.

5. At the next laboratory session, observe the streak plates. Observe where you find well-isolated colonies. Compare the colonies you observe from the mixed culture to standard colony morphologies and with the streak plates of the pure cultures; describe your results.

## Part B. Culture Transfer

1. Holding the inoculating loop between your thumb and index finger, insert the wire portion into the Bunsen burner flame, heating the entire length of the wire until it is red and glowing. Allow the wire to cool for 5–10 seconds before proceeding to the next step. Do not wave the loop in the air.

2. Using your free hand, pick up the tube containing the culture that you want to transfer and gently shake it to disperse the culture. Remove the tube cap or plug *with the free little finger of the hand holding the sterile inoculating loop* and flame the lip of the tube in the Bunsen burner flame.

3. Holding the culture tube in a slanted position, insert the sterile loop and remove a small amount of growth from the tube; a loopful is sufficient. Try not to touch the sides of the tube or sides of the lip with the loop during the removal step.

4. Flame the tube lip again, carefully replace the tube cap, and return the culture tube to the test tube rack.

5. Without flaming the loop, pick up a fresh, uninoculated test tube containing 5 ml of LB broth, remove the tube cap or plug from the tube, and flame the mouth of the tube.

6. Place the loop containing the viable inoculum into the tube containing LB broth and shake it so that the microorganisms are transferred into the tube. Remove the loop from the tube.

7. Flame the mouth of the tube, replace the cap or plug, and return the tube to the rack.

8. Flame the inoculating loop.

9. Label the freshly inoculated tube and place it in a 37°C incubator.

10. Repeat Steps 1–9, using sterile LB broth as inoculation in place of broth culture.

11. At the next laboratory session, examine the tubes and describe their appearance.

## References

These volumes are standard laboratory books for introduction to microbiological techniques.

Bradshaw, L. J. (1992). "Laboratory Microbiology," 4th Ed. Harcourt Brace, San Diego, California.

Cappuccino, J. G., and Sherman, N. (1992). "Microbiology: A Laboratory Manual," 3rd Ed. Benjamin Cummings, New York.

Harley, J. P., and Prescott, L. M. (1993). "Laboratory Exercises in Microbiology," 2nd Ed. William C. Brown, Dubuque, Iowa.

## Questions

1. How can you be certain that you have a pure culture of each organism on each plate?

2. How would you use temperature to enhance or inhibit growth of bacteria or yeast?

3. Why is it advantageous to hold the culture tube in a slanted position when inserting the loop?

4. What is the function of flaming the tube lip?

# 2

# Preparation of Culture Media

## Introduction

Any medium for the cultivation of bacteria or yeast must provide certain basic nutritional requirements, which include the following: (1) a carbon source that may also serve as an energy source, (2) water, (3) a nitrogen source, (4) a phosphorus source, (5) a sulfur source, and (6) various mineral nutrients, such as iron and magnesium. *Escherichia coli* and *Saccharomyces cerevisiae* are capable of growth on a medium consisting of a single carbon source, such as the carbohydrate glucose; a simple nitrogen source, such as ammonium chloride or ammonium sulfate; and other inorganic salts providing phosphorus, sulfur, and minerals. *Saccharomyces cerevisiae* requires some vitamins as well. This kind of medium is termed *defined* or *synthetic* because its exact chemical composition is known. For routine laboratory work, however, complex media are often employed. The basic nutrients in these media are provided by plant and animal extracts of which the exact composition is not known. For example, yeast extract and peptones (hydrolyzed protein) are the basic ingredients of the Luria–Bertani (LB) broth commonly used for *E. coli* and the yeast extract–peptone–dextrose (YEPD or YPD) broth commonly used for *S. cerevisiae*. These materials supply a variety of carbon sources; nitrogen-, phosphorus-, and sulfur-containing compounds in the form of peptides and amino acids; and a mixture of cofactors, such as vitamins. A broth medium is one in which the components are simply dissolved in distilled water. The addition of agar (a complex carbohydrate extracted from seaweed) results in a solid medium. Agar is an ideal solidifying agent for a microbiological medium because of its melting properties and because it has no nutritive value for the vast majority of bacteria and fungi. Solid agar melts at 90–100°C; liquid agar solidifies at about 42°C.

Sterilization procedures eliminate all viable microorganisms. Culture dishes, test tubes, flasks, pipets, transfer loops, and media must be free of viable microorganisms before they can be used for establishing pure cultures. The culture vessels must be sealed or capped with sterile plugs to prevent contamination. There are various ways of sterilizing the liquids, containers, and instruments used in pure culture procedures; these include exposure to elevated temperatures or radiation levels to kill microorganisms, and filtration to remove microorganisms from suspension (see Appendix 9, Sterilization Methods).

Medium preparation requires an autoclave for sterilization, which permits exposure to high temperatures for a specific period of time. Generally, a temperature of 121°C [achieved by using steam at 15 psi (lb/in$^2$)] for 20 minutes is used to heat-sterilize microbiological media.

In this exercise, you will prepare both a defined and a complex medium. You will learn how to mix the proper constituents to support the growth of *S. cerevisiae* and how to sterilize the medium.

## Reagents/Supplies

Adenine sulfate
Agar
Beakers (1 liter)
Dextrose (glucose)
Erlenmeyer flasks (500 ml and 1 liter)
Graduated cylinders (250 and 500 ml)
Heat-resistant gloves
Histidine
Lysine

Peptone
Petri plates, sterile
Plastic bags
Screw-cap bottle (100 ml)
Test tube rack
Test tubes (18 × 150 mm) with caps
Tryptophan
Uracil
Yeast extract
Yeast nitrogen base (YNB) with ammonium sulfate, without amino
    acids (Difco)

## Equipment

Autoclave
Filter sterilization unit (for 100-ml volume)
Incubator at 37°C
Magnetic stirrer
pH meter
Top-loading balance
Water bath (50°C)

### *Instructor's Note* _____

Proper budgeting of time is essential for completion of this exercise
within the laboratory period. You may wish to divide preparation
of various media among groups of students. However, we have found
it important for the students to prepare their own media rather than
having this demonstrated to them.

## Procedure

### Part A. Preparation of YEPD, a Complex Medium for Yeast [Yeast Extract (1%, w/v), Peptone (2%, w/v), Dextrose (2%, w/v)]

1. Add 500 ml of distilled water to a 1-liter Erlenmeyer flask. Weigh
   out 5 g of yeast extract, 10 g of peptone, and 10 g of glucose

(dextrose) on a top-loading balance; dissolve these in distilled water by stirring with a magnetic stirrer. (*Note:* Heating may be necessary.)

2. Divide this broth solution into two equal parts by adding 250 ml each to two 500-ml Erlenmeyer flasks:

   Flask A: Make no further additions.
   Flask B: Add 5 g of agar. Swirl to disperse the agar.

3. Dispense the broth medium (Flask A) into 15 test tubes, adding 5 ml of broth to each tube. At this time, or during the autoclaving period, use of the pH meter may be demonstrated (see Exercise 2A in Appendix 1).

4. Cap each test tube (do not tighten) and put aside for autoclaving later. Place Flask B and Flask A (with 175 ml of YEPD remaining) into the autoclave for autoclaving as well.

5. To demonstrate the necessity of the sterilization step, do not autoclave two of the test tubes; simply label and leave them on the benchtop until the next laboratory period. *Before starting Step 6, make up YNB medium components (see Part B, below).*

6. Load the autoclave with all solutions from Parts A and B that require autoclaving, and close the autoclave door. Most autoclaves have an automatic cycle; set the timer for 20 minutes on the slow exhaust mode. Start the autoclave cycle by pushing the start button in the automatic cycle mode or in the manual mode by turning the selection lever to the fill position. The sterilization cycle involves filling the jacket, allowing steam to enter the chamber, holding the temperature for the amount of time that you have set, and venting the chamber. In the automatic mode, the chamber will begin to fill with steam after the jacket pressure reaches 15–20 psi; in the manual mode, you must move the selection lever to the fill chamber position when the jacket pressure reaches this level. In manual operation, you must move the selection lever to the vent position after the chamber has been at 121°C for 20 minutes. Slow venting is required to prevent liquids from boiling.

Only after complete venting of the chamber can you open the autoclave and remove your material. Autoclaving for 20 minutes actually takes about 40 minutes when you include the time required for heating and venting the chamber. When removing material from the autoclave, use heat-resistant gloves: the material is still hot!

7. After removal from the autoclave, allow the broth tubes and 175 ml of YEPD (Flask A) to cool before tightening the caps and storing for use in later exercises.

8. After the medium in Flask B has been sterilized, place the flask in a 50°C water bath and equilibrate for 30 minutes. (The purpose of the cooling step is to prevent excessive water condensation on the Petri dish lid due to evaporation from agar at temperatures >50°C and to prevent solidification of agar at <50°C.) To dispense the medium into sterile Petri plates, flame the mouth of the flask and, while carefully lifting the lid of a Petri plate, pour about 20 ml of agar into a plate (enough medium to cover the bottom of the plate). Replace the lid and continue filling additional plates until all of the medium is dispensed. Never completely remove the lids of the Petri plates or the plates will become contaminated with bacteria and fungi from the air. Work quickly to minimize contamination, but carefully to prevent accidents. You should periodically reflame the mouth of the flask to reduce contamination. Allow the agar plates to cool. After the agar has solidified, label the plates with your name, date, and medium, and leave at room temperature or incubate at 37°C overnight to allow the agar to dry and to detect contaminated plates. Store the plates at room temperature or at 4°C in plastic bags for use in a later laboratory experiment.

## Part B. Preparation of YNB Medium, a Defined Medium for *Saccharomyces cerevisiae*

1. Add 6.7 g of yeast nitrogen base (YNB) with ammonium sulfate, without amino acids (Difco) to 100 ml of distilled water. The

suspension may require slight heating to dissolve. Filter sterilize for a 10× stock solution. Store in a refrigerator.

2. Prepare 100-ml stock solutions at 30 mg/ml of lysine, tryptophan, and histidine. Also prepare 100-ml solutions of uracil and adenine sulfate (each 2 mg/ml). All can be autoclaved except tryptophan and adenine sulfate, which should be filter sterilized. (See Appendix 10, Preparation of Stock Solutions for Culture Media.)

3. Prepare a 100-ml 20% (w/v) glucose solution, which should be autoclaved.

4. Prepare 100 ml of a 4% (w/v) agar solution. Autoclave. Place the molten agar at 55°C to equilibrate for pouring of plates on addition of prewarmed YNB, as described below.

5. Prepare minimal medium with YNB, supplements (amino acids, bases), and glucose (Table 2.1). For agar plates, add 4% agar as indicated below. Be sure to temperature equilibrate medium components to 55°C before addition of agar. Mix well, being careful not to introduce excess bubbles. Pour the plates.

**Table 2.1   Preparation of YNB Medium**

| Component | Medium | |
|---|---|---|
| | Liquid (ml) | Plates (ml) |
| YNB (10×) | 10 | 10 |
| Agar (4%, w/v) | — | 50 |
| Lysine (30 mg/ml) | 0.1 | 0.1 |
| Tryptophan (30 mg/ml) | 0.1 | 0.1 |
| Histidine (30 mg/ml) | 0.1 | 0.1 |
| Adenine sulfate (2 mg/ml) | 1.5 | 1.5 |
| Uracil (2 mg/ml) | 1.5 | 1.5 |
| Glucose (20%, w/v) | 10 | 10 |
| Distilled water | 76.7 | 26.7 |

# References

Kaiser, C., Michaelis, S., and Mitchell, A. (1994). "Methods in Yeast Genetics." Cold Spring Harbor Laboratory, Cold Spring Harbor, New York. [This is an excellent source for most of the basic methods used for yeast genetics and physiology.]

Sambrook, T., Fritsch, E. F., and Maniatis, T. (1989). "Molecular Cloning: A Laboratory Manual," 2nd Ed. Cold Spring Harbor Laboratory, Cold Spring Harbor, New York. [This was the first and remains among the most useful manuals for molecular biological techniques.]

# Questions

1. Why is the requirement for amino acids much lower than that for glucose in defined medium?

2. Explain the usage of both defined and undefined media for the growth of bacteria and yeast.

3. Why were lysine, adenine, and histidine added to the YNB medium?

**Exercise** _____

# 3

# The Growth Curve

## Introduction

Knowledge of the growth characteristics of an organism is essential to biotechnology for achieving reproducible transformation efficiency and for obtaining reproducible plasmid and recombinant protein yields. To obtain uniform, balanced growth, a culture is harvested in the logarithmic or exponential growth phase, where the growth rate and composition of each cell in the population are relatively identical (see Appendix 11, Growth in Liquid Medium, for a full description of the growth curve).

Growth varies in different media, depending on the nutrient level and aeration of the culture. Vigorous shaking is necessary to maintain sufficient dissolved oxygen throughout the culture to support growth. Even so, growth in a minimal medium compared to a rich medium will lengthen the doubling time for *Saccharomyces cerevisiae* and *Escherichia coli* (Table 3.1). This difference is due to the time and energy the bacteria must spend synthesizing metabolites otherwise supplied in complex (rich) medium. Thus, the onset and duration of log phase will vary and must be established for the particular medium and strain used. Construction of a growth curve including lag, log, stationary, and death phases will enable us to establish these parameters.

In this exercise, three methods will be used to generate data points for constructing a growth curve for *S. cerevisiae* (YNN281) grown in both complex (YEPD) and minimal (YNB) media.

*25*

**Table 3.1  Comparison of Doubling Times in Minimal and Rich Media**

| Organism | Medium | Doubling time (minutes) |
|----------|--------|-------------------------|
| *E. coli* (strain LE392) | M9 (minimal) | 50–60 |
|  | LB (rich) | 20–30 |
| *S. cerevisiae* (strain YNN281) | YNB (minimal) | 200 |
|  | YEPD (rich) | 90–100 |

1. Colony counts on agar plates, which measure only *viable* cells (see Appendix 12, Determination of Viable Cells).

2. Change in optical density of culture, which measures *cell mass* (see Appendix 13, Determination of Cell Mass).

3. Direct hemocytometer counts, which measure *cell number* (see Appendix 14, Determination of Cell Number).

Cells from an overnight culture (typically 16–18 hours) are generally in the stationary phase of the growth curve. If the culture is inoculated into fresh medium, the cells will be induced to divide and reenter the log phase (see Appendix 11, Growth in Liquid Medium, Inoculation and Subculture). We will use such a log-phase culture to make measurements.

# Reagents/Supplies

Coverslips

Culture of *Saccharomyces cerevisiae,* strain YNN281 (*MATa, ura3-52, trp1-Δ, his3-200, lys2-801, ade2-1*) (see Appendix 7, List of Cultures, and Appendix 15, Nomenclature of Strains)

Cuvettes

Dilution blanks, sterile: sterile water in test tubes (16 × 150 mm)

Ethanol (70%, v/v)

Pasteur pipets/bulbs, sterile (9 in.)

Pipets, sterile (1, 5, and 10 ml)
Spreaders (glass "hockey sticks")
YEPD agar plates
YEPD broth (50 ml) in 250-ml Erlenmeyer flask (for overnight)
YEPD broth (250 ml) (see Exercise 2) in a 1-liter Erlenmeyer flask
YNB agar plates
YNB broth (50 ml) in 250-ml Erlenmeyer flask (for overnight)
YNB broth (250 ml) (see Exercise 2) in a 1-liter Erlenmeyer flask

## Equipment

Bunsen burners
Hemocytometer
Incubator at 30°C
Microscope
Rotatry shaker at 30°C
Spectrophotometer
Vortex

### Instructor's Notes

1. Inoculate *S. cerevisiae* into 50 ml of both YEPD and YNB media.
2. Incubate with vigorous shaking overnight at 30°C. The culture should approximate $5 \times 10^7$ cells/ml by this time.
3. Dilute overnight cultures into 250 ml of their respective media to ~$5 \times 10^3$ cells/ml and incubate at 30°C with shaking.
4. Remove a 5-ml portion of cells from each culture after 0, 4, 8, 12, 16, 20, 24, 30, and 36 hours of incubation and refrigerate until class time. (The exercise is done this way because continuous monitoring of a culture over 36 hours by the class is not practical.)

5. Draw a chart on the board to be filled in with class data:

| Time point (hours) | $A_{600}$ | | Hemocytometer count | | Plate count | |
|---|---|---|---|---|---|---|
| | YNB | YEPD | YNB | YEPD | YNB | YEPD |
| 0 | | | | | | |
| 4 | | | | | | |
| 8 | | | | | | |
| 12 | | | | | | |
| 16 | | | | | | |
| 20 | | | | | | |
| 24 | | | | | | |
| 30 | | | | | | |
| 36 | | | | | | |

## Procedure

### Day 1

Each pair of students will make measurements on an assigned time point.

1. Vortex the cell suspension and aseptically remove a small amount with a sterile Pasteur pipet.

2. Charge the hemocytometer chambers with a drop of culture and count the cells microscopically (see Appendix 14, Determination of Cell Number). The density of the sample may require dilution to make the cells countable by this method. Generally, 200–500 cells in a chamber is convenient.

3. On the basis of the hemocytometer count, dilute the suspension aseptically, using sterile pipets and dilution blanks of sterile distilled water. Yeast are stable osmotically in water, owing to their thick cell walls. Bacteria must be diluted in medium or salt solution. The final tube should contian 500–2000 cells/ml.

4. Plate 0.1 ml of each of these suspensions in duplicate onto Petri plates containing either YEPD or YNB. Spread the cells with hockey stick spreaders sterilized by dipping them in ethanol and flaming them in a Bunsen burner (see Appendix 12).

5. Incubate inverted plates at 30°C and count colonies in 48 hours.

6. Dispense 1.5 ml of the original cell suspension into cuvettes.

7. Read the $A_{600}$ of the cell suspension against YEPD and YNB blanks for respective cultures. See Exercise 3A (Appendix 1) for a description of how to use a spectrophotometer.

## Day 3

1. Fill in the chart on the board with your data.

2. Plot the data of the entire class for insertion into your notebook:

   a. Cell number versus time (plot on linear and semilog paper)

   b. Plate counts (viable cells per milliliter) versus time (plot on semilog paper)

   c. $A_{600}$ versus time (plot on linear and semilog paper)

   d. $A_{600}$ versus cell number (plot on linear and semilog paper)

## References

Ingraham, J. S., Maaløe, O., and Neidhardt, F. C. (1983). "Growth of the Bacterial Cell." Sinauer Associates, Sunderland, Massachusetts. [Using *E. coli* as the model organism, the authors present results of structural, biochemical, genetic, and physiological studies in the context of cell growth and division.]

Prescott, D. M. (1975). "Methods in Cell Biology," Vol. XII. Yeast Cells. Academic Press, New York. [A compendium of information and techniques on yeast growth and physiology.]

## Questions

1. What are three methods for monitoring cell growth, and what are the respective parameters measured by each of these methods?

2. Calculate the inoculum size (cells per milliliter) for a 5 P.M. inoculation of S. *cerevisiae* YNN281 in YEPD to harvest cells at a density of $3 \times 10^8$ cells/ml at 9 A.M. the next morning.

# 4

# Isolation of Plasmid DNA from *Escherichia coli:* The Mini-Prep

## Introduction

The mini-prep is a quick method for isolating small amounts of plasmid DNA ($\sim$1 $\mu$g of DNA per milliliter of bacterial culture) from a transformed host. The mini-prep is much less labor intensive than large-scale (or maxi-prep) protocols, which utilize either a cesium chloride gradient or column chromatography for isolating milligram amounts of plasmid DNA. The mini-prep allows for rapid screening of transformants by subsequent restriction digestion of the DNA isolated. When a clone of interest is found, large amounts of DNA can then be isolated by a maxi-prep.

Various methods have been developed for rapid, small-scale plasmid isolation. Most notable are the alkaline lysis method (Birnboim and Doly, 1979) and the rapid boiling procedure (Holmes and Quigley, 1981). In this exercise, the latter procedure will be used as an introduction to recombinant DNA technology. We will introduce the alkaline lysis mini-prep method as part of another exercise later in this book (Exercise 12). Recombinant DNA methodology advances so rapidly that new technologies appear frequently. For example, several rapid procedures for the mini-prep isolation of plasmid DNA based on chromatographic purification are now available commercially from various manufacturers (Qiagen, Pharmacia, 5 prime → 3 prime, Promega). The variety of available protocols enables the molecular biologist to choose applications suitable to experimental design and budget limitations.

In many cases, plasmid DNA will be obtained as carried in a bacterial host. In this exercise, the plasmid pRY121 [see Exercise 8, Figure 8.1, for a diagram (map) of this plasmid] is carried in *E. coli* strain LE392. Selective pressure for maintenance of the plasmid is conferred by growth in ampicillin, a penicillin derivative. pRY121 contains a gene encoding β-lactamase, a periplasmic enzyme that cleaves ampicillin, thereby yielding cells resistant to this antibiotic.

A successful mini-prep leads to rapid DNA isolation for subsequent verification of plasmid identity and is an essential technique for the molecular biologist. Special care must be taken in the use of glassware for DNA or RNA to avoid contamination by hydrolytic enzymes (see Appendix 16, Glassware and Plasticware).

## Reagents/Supplies

Ampicillin (50-mg/ml stock solution in $H_2O$; filter sterilize and store at $-20°C$)

*Escherichia coli* LE392, genotype: *F⁻hsdR514* ($r⁻m⁻$) *supE44 supF58 lacY1 galK2 galT22 metB1 trpR55* $\lambda⁻$ (*pRY121*) (see Appendix 7, List of Cultures, and Appendix 15, Nomenclature of Strains)

Isopropanol (2-propanol)

Lysis buffer [8% (w/v) sucrose, 0.5% (v/v) Triton X-100, 50 mM EDTA (pH 8.0), 10 mM Tris-Cl (pH 8.0) (see Appendix 17, Preparation of Tris and EDTA)

Lysozyme [10 mg/ml in 10 mM Tris-Cl (pH 8.0)]

Microcentrifuge tubes, sterile (1.5 ml)

Micropipettors

Parafilm

Pasteur pipets

RNase A (DNase free) [10 mg/ml in 10 m$M$ Tris-Cl (pH 7.5), 15 m$M$ NaCl. Heat to 100°C for 15 minutes, cool slowly to room temperature; store at −20°C]. RNase may also be purchased ready to use from 5 prime → 3 prime, Inc.

Sodium acetate (2.5 $M$)

Styrofoam float

TE buffer [10 m$M$ Tris-HCl, 1 m$M$ EDTA (pH 8.0)]

Toothpicks

## Equipment

Aspirator

Boiling water bath, prepared just prior to use

Microcentrifuge at room temperature

Microcentrifuge at 4°C

Vacuum desiccator or centrifuge (optional)

Vortex

Water bath or shaking incubator at 37°C

### Instructor's Note

Grow 5-ml cultures of *E. coli* LE392 (pRY121) in LB medium (Luria–Bertani broth: Tryptone, 10 g/liter; yeast extract, 5 g/liter; and NaCl, 10 g/liter; pH adjusted to 7.5 dropwise with NaOH) with ampicillin (50 $\mu$g/ml) overnight (about 18 hours before class) at 37°C with vigorous aeration by shaking at ~200 rpm.

## Procedure

1. Prepare Tris and EDTA stock solutions as described in Appendix 17. Make TE buffer, lysis buffer, and lysozyme solutions (see above, Reagents/Supplies) using the Tris and EDTA stocks.

2. Aseptically transfer 1.5 ml of the overnight culture of *E. coli* LE392 into a sterile 1.5 ml microcentrifuge tube.

3. Centrifuge at top speed (~14,000 rpm) for 1 minute in microcentrifuge at room temperature.

4. Aspirate and discard spent medium and add another 1.5 ml of culture to pellet.

5. Centrifuge, discard spent medium, and repeat the procedure with another 1.5 ml of culture. At the last aspriation, leave the pellet as dry as possible but be sure not to aspirate the cells.

6. Resuspend the cell pellet by vortexing in 0.35 ml lysis buffer. This buffer is designed to weaken the outer membrane of the cell under conditions of osmotic stabilization, providing access to lysozyme to the peptidoglycan layer.

7. Add 25 $\mu$l of fresh lysozyme solution and vortex for 3 seconds. This solution should be prepared before starting the procedure and should be stored on ice.

8. Place the tube, covered with Parafilm to prevent popping of tube cover, in a Styrofoam float in a boiling water bath and boil for 40 seconds. This step serves to inactivate lysozyme.

9. Centrifuge in the microcentrifuge for 10 minutes at top speed at room temperature to pellet the remaining cell envelope with its associated chromosomal DNA.

10. With a toothpick, remove and discard the white, gelatinous pellet from the tube.

11. To the remaining supernatant, add 40 $\mu$l of 2.5 M sodium acetate and 420 $\mu$l of isopropanol to precipitate DNA. Vortex to mix and hold the solution for 20 minutes at room temperature to facilitate precipitation.

12. Centrifuge in the microcentrifuge at top speed for 15 minutes at 4°C.

13. Remove and save the supernatant until DNA precipitation is verified (Exercise 5). Dry the pellet by inverting the tube on clean, absorbent paper and letting it stand for about 20 minutes. Drying may be hastened by use of a vacuum desiccator or vacuum centri-

fuge. Care must be taken with a vacuum desiccator to avoid loss of sample.

14. Resuspend the pellet in 50 $\mu$l of TE buffer containing RNase (DNase-free; final concentration in TE of 100 $\mu$g/ml). Incubate for 10 minutes in a water bath at 37°C.

15. At this point, the sample should be stored at 4°C until further use.

## References

Birnboim, H. C., and Doly, J. (1979). A rapid alkaline extraction procedure for screening recombinant plasmid DNA. *Nucleic Acids Res.* **7**, 1513–1518. [See Exercise 6 for a description of this technique and Exercise 12 for its application in mini-preps.]

Holmes, D. S., and Quigley, M. (1981). A rapid boiling method for the preparation of bacterial plasmids. *Anal. Biochem.* **114,** 193–197. [This paper introduces the mini-prep procedure used in this exercise and has subsequently been adopted by many investigators.]

Murray, N. E., Brammer, W. J., and Murray, K. (1977). Lambdoid phages that simplify the recovery of *in vitro* recombinants. *Mol. Gen. Genet.* **150,** 53–61. [A description of the host *E. coli* strain LE392 is included in this paper.]

West, R. W., Jr., Yocum, R. R., and Ptashne, M. (1984). *Saccharomyces cerevisiae GAL1–GAL10* divergent promoter region: Location and function of the upstream activating sequence UAS$_G$. *Mol. Cell. Biol.* **4,** 2467–2478. [This reference describes the construction of plasmid vector pRY121. Refer to Exercise 8 for a map of this vector.]

## Questions

1. Explain the significance of the following reagents as applicable to the mini-prep procedure: lysozyme, sucrose, RNase, ampicillin, and isopropanol.

2. Show your calculations for preparing the following stock solutions: 20% (w/v) glucose, 0.2 $M$ EDTA, 0.75 $M$ Tris, 1 $M$ NaOH. How much of the above stock solutions would you use to prepare 100 ml of TE buffer which is 10 m$M$ Tris and 1 m$M$ EDTA?

**5**

# Purification, Concentration, and Quantitation of DNA

## Introduction

Purification of DNA from a complex mixture of cellular molecules is most readily accomplished by removal of proteins and other molecules into an organic solvent. This extraction procedure takes advantage of properties of phenol and phenol–chloroform that lead to denaturation of proteins. DNA and RNA are not soluble in the organic solvents, and thereby remain associated with aqueous phases of mixtures that contain solvents for protein extraction.

For concentration of purified DNA, the method most widely used is precipitation with ethanol. The precipitated DNA may be recovered by centrifugation and redissolved in a small amount of buffer. In this manner, DNA solutions of desired concentration can be obtained from very small amounts of DNA (even nanogram levels). Precipitation with ethanol will also remove traces of phenol and chloroform that would otherwise inhibit restriction enzymes and other enzymes used in molecular cloning. Ethanol precipitation also removes unincorporated nucleotides and oligonucleotide primers from labeled DNA or polymerase chain reaction (PCR) solutions (see Exercise 7 on the PCR of the gene encoding $\beta$-galactosidase). Two common protocols employ either sodium acetate and ethanol or ammonium acetate and ethanol. Both of these salts have been proven equally effective for recovery of DNA from small volumes.

Quantitation of nucleic acids can be carried out by several methods. Nucleic acids absorb ultraviolet light at 260 and 280 nm and bind the fluorescent dye ethidium bromide (EtBr). These physical characteristics form the bases of the most convenient and common methods used to measure the amount of DNA. Furthermore, ultraviolet (UV) light absorption can be used to assess the purity of DNA. Spectrophotometric determination is the method of choice if sufficient quantities of relatively pure DNA are to be assayed. Ethidium bromide binding is useful when only small amounts of DNA, or contaminating UV-absorbing material, are present. Unfortunately, ethidium bromide is a carcinogen; it must be handled and disposed of with caution. Nontoxic alternatives to ethidium bromide are available (e.g., Hoescht dye 33258) but often require use of a fluorimeter or are relatively expensive. This exercise demonstrates three common methods for DNA quantitation: UV absorption, ethidium bromide fluorescence, and "band-brightness" comparisons following agarose gel electrophoresis.

## PART A     PURIFICATION OF PLASMID DNA BY SOLVENT EXTRACTION

### Reagents/Supplies

Chloroform (a mixture of chloroform: isoamyl alcohol, 24 : 1, v/v)
DNA sample, mini-prep (see Exercise 4)
Ethanol (100%; store at −20°C)
Gloves
Ice
Microcentrifuge tubes (1.5 ml)
Micropipettors
Phenol (see Instructor's Note, below)
TE buffer (pH 8.0) (see Exercise 4)

### Equipment

Microcentrifuge at 4°C
Microcentrifuge at room temperature

***Instructor's Note*** _____

It is important that phenol be free of contaminants that cause DNA damage. The instructor will provide the class with purified phenol prepared according to Sambrook *et al.* 1989). Purified, redistilled phenol can be purchased from many companies, circumventing the redistillation steps. We strongly recommend using commercially available phenol with the simple additional equilibration with TE buffer according to Sambrook *et al.* (1989). Additionally, several companies now offer a ready-to-use buffered phenol solution that is convenient and stable for several months.

## Procedure

***Safety Note*** _____

When using phenol, wear gloves and work under a fume hood.

1. Mix 50 $\mu$l of mini-prep DNA sample (From Exercise 4) in a 1.5-ml microcentrifuge tube with 50 $\mu$l of TE buffer (pH 8.0) to obtain a workable volume.

2. To this mixture add 100 $\mu$l of phenol. This organic solvent serves to denature and extract protein.

3. Mix the contents by inverting the tube gently several times, until an emulsion forms. This avoids breakage of DNA that occurs by shear forces generated in vortexing and violent stirring.

4. Centrifuge in the microcentrifuge for 20 seconds at top speed at room temperature (about 14,000 rpm).

5. With a micropippettor, transfer the upper aqueous phase to a clean 1.5-ml microcentrifuge tube.

6. Repeat the extraction of the remaining lower organic phase and interphase in the original tube by adding 100 $\mu$l of TE buffer. Mix by inversion. Centrifuge as in Step 4. Collect the aqueous phase and combine with the first aqueous phase collected.

7. Extract the combined aqueous phases by adding about 100 $\mu$l of phenol and 100 $\mu$l of chloroform. The chloroform also serves to extract and denature protein. Mix by inversion, centrifuge as in Step 4, and transfer the upper phase to a clean 1.5 ml microcentrifuge tube.

8. Extract the upper, aqueous phase from Step 7 with 200 $\mu$l chloroform only. This final chloroform extraction serves to remove residual phenol from the DNA-containing aqueous phase. Mix, centrifuge as in Step 4, and collect the upper phase that contains purified DNA, which should be stored at 4°C, or on ice if proceeding directly to Part B of this exercise.

## PART B     CONCENTRATION OF PLASMID DNA

### Reagents/Supplies

Ammonium acetate (10 $M$)
DNA sample from Part A of this exercise
Ethanol (100% at −20°C)
Ice
TE buffer (pH 8.0)

### Equipment

Microcentrifuge at room temperature
Vacuum centrifuge or desiccator (optional)

### Procedure

1. Adjust the concentration of the DNA purified from Step 8, Part A, by adding 10 $M$ ammonium acetate to make a final solution that contains 2 $M$ ammonium acetate. This is done by estimating

the volume of the DNA solution that is in TE buffer (pH 8.0). This is necessary to facilitate precipitation of DNA by ethanol.

2. Add 2 volumes of cold 100% ethanol. Mix and store on ice for 10 minutes. (If the DNA is smaller than 1 kb or present at a concentration less than 100 ng/ml, the solution should be stored at $-70°C$ for about 4 hours. For DNA less than 0.2 kb in size, the addition of 0.01 $M$ $MgCl_2$ improves recovery.)

3. Centrifuge at 4°C for 10 minutes in a microcentrifuge at top speed ($\sim$14,000 rpm).

4. Discard the supernatant. Invert the tubes on a layer of absorbent paper to allow drainage of ethanol. Solvent traces can be removed in a vacuum desiccator or vacuum centrifuge.

5. Dissolve the pellet in 25 $\mu$l of TE buffer (pH 8.0). Rinse the tube walls with buffer to ensure dissolution of DNA. Heating to 37°C for at least 5 minutes may help solubilize the DNA as well.

6. Store the DNA solution at 4°C. This purified mini-prep DNA will be used in Part C for quantitation of DNA.

# PART C    QUANTITATION OF DNA

## Reagents/Supplies

### *Safety Note* _____
Ethidium bromide is highly toxic. Wear gloves, goggles, and a laboratory coat when working with EtBr (see Instructor's Note, below).

DNA sample from Part B of this exercise

DNA standards [λ DNA digested with *Hin*dIII, 20 to 1 $\mu$g/ml in TE buffer (pH 8.0)]

Ethidium bromide, stock solution (10 mg/ml) (see Instructor's Note, below)

Ethidium bromide waste container

Goggles or face shield, UV-blocking
Plasmid DNA sample from Exercise 4 or Exercise 6
Plastic wrap
Polaroid film (type 667)
Quartz cuvette, 1-ml capacity (or smaller)
Sterile water
TE buffer (see Exercise 4)

## Equipment

Polaroid camera, type MP-4 or equivalent
Spectrophotometer
Ultraviolet transilluminator

### *Instructor's Note* _____

To prepare the EtBr stock solution, add 1 g of EtBr to a 100-ml graduated cylinder. Add 1 ml of 95% (v/v) ethanol and use a magnetic stirrer to mix. The EtBr will dissolve in about 5 minutes. Bring the volume to 100 ml with distilled water for a 10-mg/ml stock solution. Ethidium bromide is soluble in ethanol and sparingly soluble in water. Ethidium bromide is light sensitive and should be stored in a brown or foil-wrapped bottle. It is highly toxic. Wear gloves, goggles, and a laboratory coat.

### *Note* _____

**Inactivation of Ethidium Bromide**   Ethidium bromide is recognized as a potent mutagen and must be disposed of in a responsible manner. We list two alternatives for EtBr waste removal:

1.  Ethidium bromide decontamination from liquid solutions can be performed using a filtration/extraction device available at a reasonable price from Schleicher & Schuell. Supelco, Inc., also offers an extraction device for EtBr decontamination.
2.  A more tedious, but effective, alternative involves the following: For every 100 ml of EtBr, add 25 ml of freshly prepared 5%

hypophosphorous acid solution and 12 ml of 0.5 $M$ fresh sodium nitrite. The final pH of solution should be <3.0. Stir briefly and let stand for 20 hours. Neutralize with an excess of 1 $M$ sodium bicarbonate and discard. In contrast to common practice, inactivation by sodium hypochlorite (bleach) is not effective in eliminating mutagenicity (Lunn and Sandstone, 1987).

# Procedure

## *Determination by Ultraviolet Absorption*

A pure solution of double-stranded DNA at 50 $\mu$g/ml has an optical density of 1.0 at 260 nm and an $OD_{260}/OD_{280}$ ratio of 1.8. Contamination with protein or phenol will give $OD_{260}/OD_{280}$ values significantly less than 1.8 and contamination with RNA gives a ratio greater than 1.8. For pure RNA, $OD_{260}/OD_{280} = 2.0$. Thus, the $OD_{260}/OD_{280}$ value is obtained first, and, if it approaches 1.8, an accurate estimation of DNA concentration can be determined from the absorption at 260 nm.

1. Use DNA standards to construct a standard curve of $OD_{260}$ versus DNA concentration. (*Note:* See Appendix 1, Exercise 3A, Use of the Spectrophotometer.)

2. Dilute the mini-prep DNA (from Part B) 1:100 in TE buffer (pH 8) and measure the $OD_{260}$ and $OD_{280}$ with a spectrophotometer.

3. Determine the purity and calculate the concentration of the purified mini-prep DNA.

## *Determination by Ethidium Bromide Fluorescence*

If the DNA concentration is less than 250 ng/ml, spectrophotometric assay is not possible. A more sensitive assay takes advantage of the fluorescence of a complex formed between DNA and EtBr. Ethidium bromide binds to DNA by intercalation so that the total fluorescence is proportional to the DNA mass.

1. Stretch commercial plastic wrap over an ultraviolet transilluminator surface.

2. In an orderly array, using a micropipettor, spot samples of DNA (1–5 $\mu$l) and 2 $\mu$l of a series of DNA standards (0.5 to 20 $\mu$g/ml, prepared by serial dilution) onto the plastic.

3. To each spot add an equal volume of TE buffer (pH 8.0) containing ethidium bromide (2 $\mu$g/ml). Mix with a micropipettor.

4. Illuminate the spots with a short-wave ultraviolet light and photograph with a Polaroid camera. You should be able to estimate the concentration of the DNA by comparison to the fluorescence intensities of the known standards.

### Safety Note

Ultraviolet light is hazardous and exposure should be minimized. Wear goggles or a face shield.

## Determination of "Band-Brightness" Comparison Following Agarose Gel Electrophoresis

### Instructor's Note

This method for DNA quantitation requires that the plasmid sample be digested with a restriction endonuclease and electrophoresed on an agarose gel next to a known quantity of a DNA molecular weight standard. Therefore, it is best to perform this part of the exercise in conjunction with Exercise 8, as a means of quantifying the plasmid pRY121 DNA isolated in Exercise 6 by column chromatography or in Exercise 4, The Mini-Prep.

A rough estimation of DNA quantity can be obtained by visually comparing the fluorescence of a specific "band" or DNA fragment that has been subjected to agarose gel electrophoresis and stained with EtBr to the fluorescence of another standard "band" (of known size and amount).

1. Digest 1 $\mu$l of plasmid DNA (pRY121 mini-prep from Exercise 4 or maxi-prep from Exercise 6) as outlined in Exercise 8. [*Note:* To simplify the quantitation of total plasmid DNA in this 1-$\mu$l sample, digest the DNA with an enzyme that linearizes the plasmid, thereby generating one "band" on the subsequent agarose gel (see Exercise 8 for a map of pRY121).]

2. Following digestion of the plasmid sample, load the entire volume of the enzymatic reaction onto an agarose gel after adding an appropriate quantity of 6× loading dye. [For example, if you performed the digest in a total volume of 20 $\mu$l, you should add 4 $\mu$l of 6× dye to the reaction tube, then load the entire 24 $\mu$l.]

3. Mix 1 $\mu$l (500 ng) of *Hind*III-digested bacteriophage $\lambda$ DNA (500 ng/$\mu$l stock) with 9 $\mu$l of sterile water and add 2 $\mu$l of 6× gel loading dye. Load this entire mixture as a molecular weight standard onto the agarose gel.

4. Following electrophoresis (as outlined in Exercise 8), stain the gel with EtBr (final concentration, 0.5 $\mu$l/ml) for 5 minutes. Decant the stain into an EtBr waste container and destain the gel in distilled water for 10 minutes. View the gel by means of a UV transilluminator and take a photograph.

### Safety Note

Do not attempt to estimate approximate band fluorescence by direct visualization of the UV-illuminated DNA. Take a photograph to avoid potential eye damage or skin burn caused by prolonged exposure to UV light.

5. Visually determine which of the fluorescing $\lambda$ *Hind*III standard bands corresponds best to the intensity of the fluorescing plasmid band you wish to quantitate. Note the size of this $\lambda$ band.

6. Because it can be assumed that the amount of fluorescence in both bands is due to an approximately equivalent quantity of DNA,

this information can be used to estimate the amount of DNA loaded in the plasmid sample by virtue of knowing the percentage of the total quantity of standard DNA represented in the λ band.

For example, if you determine that the brightness of the plasmid band is similar in brightness to the 4.4-kb band of λ *Hind*III, then calculate what percentage of the total λ DNA that was loaded is represented by the 4.4-kb band.

Since complete, undigested bacteriophage λ DNA is 48.5 kb in total size, the 4.4-kb band represents about 9% [4.4/48.5 = 0.09(× 100)] of the total λ DNA loaded in the sample. Nine percent of the 500 ng loaded in Step 3 is equal to 45 ng; therefore the amount of DNA loaded in the plasmid lane would also be approximately 45 ng. If only 1 $\mu$l of plasmid DNA was used in the restriction digest and subsequently loaded onto the gel, the concentration of plasmid in the original sample would be 45 ng/$\mu$l.

## References

Crouse, J., and Amorese, D. (1987). Ethanol precipitation: Ammonium acetate as an alternative to sodium acetate. *FOCUS* (Life Technologies, Inc.) **9**(2), 3. [A comparison of various methods and factors influencing DNA precipitation.]

Gallagher, S. (1994). Quantitation of DNA and RNA with absorption and fluorescense spectroscopy. In "Current Protocols in Molecular Biology" (Ausubel, F. M., *et al.*, eds.). John Wiley & Sons, New York, pp. A.3.D.1–A.3.D.8. [This reference is an appendix within an excellent molecular biology technical text. It refers to the various quantitation methods and associated parameters.]

Lunn, E. G., and Sandstone, F. (1987). Ethidium bromide: Destruction and decontamination of solutions. *Anal. Biochem.* **162**, 453–458.

Sambrook, J., Fritsch, E. F., and Maniatis, T. (1989). "Molecular Cloning: A Laboratory Manual," 2nd Ed. Cold Spring Harbor

Laboratory, Cold Spring Harbor, New York. [In the three-volume set, commonly referred to as the "Bible" of molecular biology, several methods for DNA purification and quantitation are outlined in detail.]

## Questions

1. Calculate the amount of DNA in a 150-$\mu$l sample with an $OD_{260}$ reading of 0.012.

2. What steps would be taken if your DNA preparation had an $OD_{260}$/$OD_{280}$ ratio of 1.49 or 1.98?

3. In Part B of this exercise, how manyfold was the DNA concentrated?

# Large-Scale Isolation of Plasmid DNA by Column Chromatography

## Introduction

Large-scale plasmid preparations, often referred to as "maxi-preps" because they are designed for the isolation of milligram quantities of plasmid (versus only a few micrograms in a "mini-prep"), can be performed using a variety of procedures. All these methods require growth of microorganisms that contain the plasmid, harvesting and lysis of cells, and purification of plasmid DNA by separating it from proteins and chromosomal DNA.

In general, bacterial growth and cell lysis are achieved in a similar manner, irrespective of the protocol used. High quantities of plasmid are obtained by allowing cells to reach near saturable levels of growth prior to harvesting. Subsequent lysis using a mixture of mild detergent–alkali solution results in the release of the compact, supercoiled plasmid molecules into solution, whereas most of the larger chromosomal DNA (which, in bacteria, is attached to the cell membrane) remains associated with the cellular debris following centrifugation. Because they are covalently closed circles, the double-stranded plasmid DNA molecules are not denatured by the alkali.

Following lysis, purification of plasmid DNA away from host proteins and the remaining chromosomal DNA contaminants can be accomplished by a variety of methods. The traditional procedure for large-scale plasmid purification involves CsCl gradient centrifugation (see Appendix 1, Exercise 6A). While this method yields plasmid DNA in high quantity and quality, it is unfortunately time consuming and utilizes carcinogens and organic solvents. Advances in the development of novel DNA-affinity matrices now allow for the purification of milligram quantities of plasmid DNA by safe and effective column chromatography methods. In particular, this exercise demonstrates the use of a solid-phase anion-exchange resin (manufactured by Qiagen, Inc., Chatsworth, CA) that selectively enables the separation of nucleic acids from other cellular contaminants by a gravity-based chromatography procedure. This resin has a large pore size (approximately 100 $\mu$m) and is composed of a hydrophilic surface coated with a high density of charged diethylaminoethyl (DEAE) groups. These attributes allow for a broad separation range (from 0.1 to 1.6 $M$ salt) that enhances the isolation of plasmid DNA from complex mixtures also containing cellular proteins, RNA, and chromosomal DNA. This entire procedure can be performed rapidly (in a few hours) and the purity of plasmid obtained in this procedure is equivalent to, or greater than, that resulting from the use of gradient centrifugation with CsCl.

## Reagents/Supplies

Ampicillin, 50-mg/ml stock solution, filter sterilized

Buffer P1 (resuspension buffer): 50 m$M$ Tris-Cl, 10 m$M$ EDTA (pH 8.0). *Note:* Add RNase A (supplied as a powder in this kit) to Buffer P1 at a final concentration of 100 $\mu$g/ml; store at 4°C; solution is stable for ~6 months

Buffer P2 (lysis buffer): 200 m$M$ NaOH, 1% (w/v) sodium dodecyl sulfate (SDS). Store at room temperature

Buffer P3 (neutralization buffer): 3.0 $M$ potassium acetate (pH 5.5). Store at 4°C, chill on ice before use

Buffer QBT (column equilibration buffer): 750 m$M$ NaCl, 50 m$M$ morpholinepropanesulfonic acid (MOPS), 15% (v/v) ethanol (pH 7.0), 0.15% (v/v) Triton X-100. Store at room temperature

Buffer QC (wash buffer, pH 7.5): 1.0 $M$ NaCl, 50 m$M$ MOPS, 15% (v/v) ethanol. Store at room temperature

Buffer QF (elution buffer, pH 8.5): 1.25 $M$ NaCl, 50 m$M$ Tris-Cl, 15% (v/v) ethanol. Store at room temperature

Centrifuge bottles, 500 ml, polycarbonate (Nalgene), sterilized

Centrifuge tube racks

Centrifuge tubes, disposable, sterile (50 ml) (Falcon, type 2098)

Centrifuge tubes, 50 ml, Oak Ridge round bottom, polycarbonate (Nalgene), sterilized. Although these tubes are sold and designated as "50 ml," their actual capacity is ~35 ml

*Escherichia coli* LE392 transformed with pRY121, streak plate

Ethanol, 70% (v/v) stored at −20°C, use cold

Ice

Isopropanol, 100%

LB broth (see Exercise 3)

Pipets

Plasmid Maxi Kit (Qiagen)

Qiagen Tip 500 columns (10 per kit) and plastic tip holder

TE buffer, pH 8.0 (see Exercise 4)

## Equipment

Centrifuge (Beckman J2-21 or equivalent)

Centrifuge rotor (Beckman JA-10 or equivalent)

Centrifuge rotor (Beckman JA-20 or equivalent)

Water bath or incubator shaker, set at 37°C

### Instructor's Note

Approximately 16–18 hours before the start of class, inoculate a 2-liter flask holding 500 ml of LB broth containing ampicillin (50 μg/ml) with *E. coli* LE392 transformed with pRY121. One 500-ml culture for each group is required. Shake the cultures in a water bath or incubator shaker at 37°C (~200 rpm) until class time. The culture should be at a cell density of $OD_{600}$ = 1.0–1.5.

## Procedure

1. Harvest the *E. coli* cells by centrifugation in two sterile 500-ml polycarbonate bottles. Spin at 4°C for 10 minutes at 6000 *g* in a Beckman JA-10 rotor or equivalent.

2. Discard the supernatant by inverting the open centrifuge bottle and draining all the medium.

3. Add 10 ml of buffer P1 to one of the bottles containing a bacterial pellet and resuspend the contents by swirling repeatedly. Resuspension can be facilitated by pipetting the cell–buffer mixture repeatedly as well. Transfer this mixture to the second bacterial pellet and resuspend the combined cells. (*Note:* Buffer P1 should contain RNase.)

4. Transfer the cell suspension to a sterile 50-ml Oak Ridge polycarbonate centrifuge tube and add 10 ml of buffer P2. Cap the tube and invert it approximately 10 times to mix the contents thoroughly. *Do not vortex,* as this will result in shearing of genomic DNA. Incubate at room temperature for 5 minutes to allow for cell lysis by the alkaline solution.

5. Add 10 ml of chilled buffer P3 and mix by inverting the tube approximately five or six times immediately. Place the tube on ice for 20 minutes to allow for precipitation of cellular proteins.

6. Centrifuge the mixture at 15,000 rpm (~30,000 *g*), using a Beckman JA-20 rotor or equivalent, for 30 minutes at 4°C. Remove the supernatant immediately after centrifugation and transfer it to a fresh sterile Oak Ridge tube (50 ml). If the supernatant still contains particulate material, repeat this centrifugation step to clarify the supernatant (a 15-minute spin is sufficient).

7. Position a Qiagen Tip 500 column in a plastic tip holder and allow the holder to rest on the top of a disposable 50-ml centrifuge

tube in a test tube rack. Equilibrate the resin within the tip by pipetting 10 ml of buffer QBT into the column. Allow the buffer to drain through the column into the disposable centrifuge tube by gravity flow until the column appears empty. Empty the disposable waste collection tube following this step and replace it for subsequent waste collection. (*Note:* To save time, this step can be performed during the centrifugation in Step 6. The column is designed with a frit in it to prevent it from drying out.)

8. Apply the supernatant from Step 6 onto the column and allow it to enter by gravity flow. The nucleic acid in the supernatant will remain bound to the resin along with other contaminants at this stage of the procedure.

9. As soon as this solution has passed through the column, wash the resin by applying 10 ml of buffer QC to the column and collect the wash in a tube. When this wash has been completed, repeat with another 10 ml. Empty the waste collection tube as necessary and discard it following this step.

10. The plasmid DNA can now be eluted by placing the column on top of a fresh Oak Ridge tube (it must be held in place by hand) and adding 15 ml of buffer QF. Empty the column by force of gravity.

11. The DNA is precipitated in the Oak Ridge tube by the addition of 10.5 ml of 100% isopropanol and mixing by inversion 10 times. Centrifuge at 10,000 rpm (~15,000 $g$) at 4°C for 30 minutes using a Beckman JA-20 rotor, then carefully decant the supernatant. Using a permanent marker, mark the outside of the tube at the spot where the DNA pellet is observed (sometimes difficult to see).

12. Wash the DNA pellet by adding 15 ml of cold 70% (v/v) ethanol to the tube and gently swirl. Centrifuge at 10,000 rpm at 4°C for about 5 minutes and carefully aspirate off the supernatant. Allow the pellet to air dry for 10 minutes, then redissolve the DNA in 500 $\mu$l of TE buffer (pH 8.0). In the process of adding

the TE buffer, repeatedly pipet the buffer solution down the side of the tube marked previously in order to maximize plasmid recovery. Store the DNA solution at 4°C for future use.

### Instructor's Notes

1. The pRY121 DNA isolated should be quantitated by one of the methods described in Exercise 5. This should be performed by either the instructor or the students prior to using this plasmid DNA in Exercise 8.
2. The plasmid DNA isolated in this procedure could also be compared with respect to quantity and purity to that obtained in either Exercise 4 (the mini-prep) or Exercise 6A (see Appendix 1).

## References

Birnboim, H. C., and Doly, J. (1979). A rapid alkaline extraction procedure for screening recombinant plasmid DNA. *Nucleic Acids Res.* 7, 1513–1522. [The original citation of the alkaline lysis method of plasmid isolation. This is a landmark paper, as this method or variations of it have been adopted by most molecular biologists as the standard.]

Qiagen, Inc. (1995). "QIAGEN Product Guide." Qiagen, Inc., Chatsworth, California. [This catalog describes the procedures used in the anion-exhange resin maxi-prep purification of DNA. It also contains useful information with respect to nucleic acid separation.]

## Questions

1. How would you determine the purity of the plasmid obtained from this procedure? Name two possible contaminants and the stages in this protocol at which they should have been removed.

2. What constituents of the buffer solutions used in this procedure play the most significant role in allowing for proper nucleic acid separation and elution?

# 7

# Amplification of a *lacZ* Gene Fragment by the Polymerase Chain Reaction

## Introduction

The polymerase chain reaction (PCR) is an extremely useful technique that has revolutionized the manner by which investigators can amplify, detect, manipulate, and clone DNA fragments from a wide variety of sources. Biotechnological applications of PCR methodology include everything from improvements in basic molecular biology methods (i.e., probe generation, DNA sequencing, and targeted mutagenesis) to clinical uses in the diagnostic detection of genetic mutations, viral infections, or molecular "fingerprinting" as applied to forensic medicine.

The PCR is based on the premise that molecules of double-stranded DNA can be denatured and that specifically designed synthetic oligonucleotides, complementary to each of the two strands of the helix and lying at opposite ends of a particular region, can serve to prime *in vitro*, DNA polymerase-mediated synthesis of a double-stranded DNA region defined by the two oligonucleotides. Repeated rounds of denaturation, primer annealing, and extension yield increasing quantities of product, resulting in the exponential amplification of the originally designated sequence (see Figure 7.1). Therefore, only very small quantities of target or "template" DNA are initially required to obtain significantly large amounts by PCR. The availability of a purified heat-stable DNA polymerase from the thermophilic

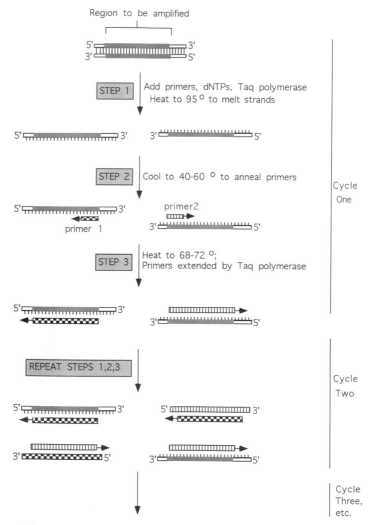

**Figure 7.1**    Amplification of a designated sequence by PCR.

bacterium *Thermus aquaticus* (termed *Taq* polymerase) greatly facilitated the enzymatic catalysis of DNA amplification, owing to its ability to withstand repeated cycles of high temperatures necessary for DNA denaturation. Genetically engineered, heat-stable DNA polymerases have been introduced that increase the fidelity of amplification over that of the *Taq* polymerase. Likewise, automation of PCR by the development of rapid temperature-cycling devices has provided a convenient means by which these reactions can be performed.

Several factors influence the fidelity and efficiency of the PCR process. The concentration of various components constituting a typical PCR is a primary determinant of successful amplification. These include the concentration of *Taq* polymerase, deoxynucleotide triphosphates, magnesium ions, template DNA, and primers in the reaction. Additionally, the temperature and times of each step within each cycle and total number of cycles can dramatically affect product yield and specificity. The term *stringency* (also see Exercise 10) refers to a concept used to describe the differential effects of temperature and salt concentration in nucleic acid hybridization and is important in discerning appropriate conditions for PCR. Conditions of high stringency involve high temperature and low salt concentrations, which lead to a decrease in the stability of DNA hybridization, whereas a low-temperature/high-salt environment is reflective of low-stringency conditions and tends to increase overall hybridization. By varying the $MgCl_2$ concentration and annealing temperature an investigator can change the stringency of hybridization during PCR and optimize amplification conditions for a specific primer pair and template. Finally, the design of oligonucleotide primers is, perhaps, the most critical parameter in PCR. A general "rule of thumb" is to utilize primers approximately 20 bases in length and containing at least 50% guanine and cytosine nucleotides in order to maximize hybridization to a given target sequence. In all, successful PCR is an empirically determined process and often requires the establishment of amplification conditions for each new template or primer pair. This exercise employs PCR to amplify a specific region of DNA within the *Escherichia coli* β-galactosidase gene, which will be subsequently analyzed in Exercise 8 by agarose gel electrophoresis.

## Reagents/Supplies

Deoxynucleotide triphosphate mix (dNTPs), 1 m$M$ mixture containing all four dNTPs. (*Note:* Individual dNTP stocks can be purchased from several manufacturers; see Appendix 20)

Ice

Magnesium chloride (MgCl$_2$), 25 m$M$

Microcentrifuge tubes, 0.5-ml capacity

Micropipettor

Mineral oil, molecular biology grade (Sigma)

Plasmid pRY121 DNA (from Exercise 6); prepare a 10-$\mu$l working stock at 2 $\mu$g/ml

Synthetic oligonucleotide primers, 10 $\mu M$ working stock in sterile water:

Primer A: 5'-GTTGTTGCAGTGCACGGCAG-3'
Primer B: 5'-GCTGGAATTCCGCCGATACTG-3'
(*Note:* Oligonucleotides may be purchased from a variety of manufacturers; see Appendix 20)

*Taq* DNA polymerase, 5 units/$\mu$l (Promega Corp. or Perkin-Elmer Corp.)

*Taq* polymerase reaction buffer, 10× (usually supplied with enzyme by manufacturer): 500 m$M$ KCl, 100 m$M$ Tris-Cl (pH 9.0), 1.0% (v/v) Triton X-100

## Equipment

DNA thermocycler or equivalent (Perkin-Elmer); alternatively, three water baths set at 94, 55, and 72°C

Heat block (optional)
Water bath (optional)

### *Instructor's Notes*

1. The cost of a DNA thermocycler may be prohibitive to some laboratories wishing to utilize PCR methodology. Although the parameters outlined in this procedure are designed for use with an automated thermocycling device, they may be empirically modified for use with less expensive methods. In their book on PCR protocols, Innis and Gelfand (1990) have published an overview of several inexpensive alternatives for automating PCR that may be of interest to those on limited budgets. Likewise, substitution of water baths for an automated thermocycler can be used as an economic alternative for performing PCR (described by Coen, 1991).

2. The oligonucleotide primers listed above are designed to specifically amplify a 562-bp DNA fragment within the open reading frame of the *lacZ* gene (toward the 3' end). The actual sequence of the entire *lacZ* gene can be accessed through the GenBank database (Accession No. V00296). The primers chosen for use in this exercise have no significance other than that they are designed to amplify part of this gene and demonstrate desirable features for PCR (i.e., the primers are specific for the target DNA to be amplified and they anneal well at 55°C to the target DNA).

## Procedure

1. Using a micropipettor, prepare a PCR reaction by pipetting the following into a 0.5-ml microcentrifuge tube:

| | |
|---|---|
| Distilled $H_2O$ | 77.0 $\mu$l |
| dNTPs, 1 m$M$ | 2.0 $\mu$l |
| *Taq* DNA polymerase reaction buffer, 10× | 10.0 $\mu$l |
| $MgCl_2$ | 6.0 $\mu$l |
| Primer A, 10 $\mu M$ | 2.0 $\mu$l |
| Primer B, 10 $\mu M$ | 2.0 $\mu$l |
| pRY121 DNA, 2 $\mu$g/ml | 0.5 $\mu$l |

2. Prepare another reaction, as described in Step 1, containing all the components listed except the pRY121 DNA. This will represent a control reaction. Place the tubes on ice.

3. Program a DNA thermocycler to run *30 cycles* with the following parameters:

Denature:   96°C for 30 seconds
Anneal:     55°C for 30 seconds
Extension:  72°C for 60 seconds

Cycling should conclude with a final extension at 72°C for 5 minutes after completion of the thirtieth cycle. This step ensures that the final product is completely polymerized.

4. After programming the thermocycler, add 0.5 μl of *Taq* DNA polymerase to each reaction. Mix the contents of each tube well by repeatedly pipetting the mixture several times, using a micropipettor.

5. Overlay each reaction mixture with 2 drops (~50 μl) of mineral oil. Cap the tubes tightly and place them into the heating block of the thermocycler. Initiate the cycling program.

6. At completion of the final extension period, place the reactions at 4°C for analysis in Exercise 8 or at −20°C if stored for longer than a few days.

## Instructor's Notes

1. It is sometimes useful, but not always necessary, to incorporate an extended precycling denaturation step of heating the reactions to 96°C for 10 minutes (in either a heat block, water bath, or thermocycler) prior to the addition of *Taq* DNA polymerase. This ensures that the template is sufficiently denatured for subsequent cycling.
2. In cases where water baths are substituted for an automated cycling apparatus, it is necessary to move the reaction tubes manually between baths set at different temperatures. As the time for reaching the desired temperature mark is greater in the absence of an

automated thermocycler, the incubation times at each step must be adjusted accordingly to allow for sufficient denaturation, annealing, and extension. Suggested incubation times are about 90 seconds, 2 minutes, and 3 minutes, respectively.

## References

Coen, D. M. (1991). Enzymatic amplification of DNA by PCR: Standard procedures and optimization. In "Current Protocols in Molecular Biology" (Ausubel, F. M., *et al.*, eds.). John Wiley & Sons, New York, pp. 15.1.1–15.1.7. [This describes a protocol for PCR that substitutes water baths for an automated thermocycler. It is also an excellent reference citing parameters that influence PCR.]

Innis, M. A., and Gelfand, D. H. (1990). Optimization of PCRs. In "PCR Protocols: A Guide to Methods and Applications" (Innis, M. A., *et al.*, eds.). Academic Press, San Diego, California, pp. 3–12. [This is an excellent chapter within a prime reference book on PCR methods. This chapter provides a basic procedure for PCR and discusses the contribution of each component in a PCR reaction required for successful amplification.)

Mullis, K. B. (1990). The unusual origin of the polymerase chain reaction. *Sci. Am.* April, pp. 56–65 [An entertaining and informative account of the discovery of PCR, written by its Nobel Laureate inventor. This article also represents a good, basic introductory level description of the technique.]

## Questions

1. Suppose your PCR reaction yielded multiple products in addition to the one you desired; without using different primers, what changes would you make to the reaction to increase the specificity?

2. If the cells of a particular individual contained a nucleotide base mutation on an allele of chromosome 17 and PCR was used to

amplify a region of DNA flanking the mutation, using DNA isolated from cells of this individual, approximately what percentage of the amplification products would contain the mutation?

3. If a PCR did not yield any product where one was expected, suggest four things you could alter to obtain a product. Explain the rationale behind each suggestion.

_____

# 8

# Restriction Digestion and Agarose Gel Electrophoresis

## Note _____

Refer to Appendix 18 (Basic Rules for Handling Enzymes).

## Introduction

Restriction endonucleases are enzymes that recognize specific nucleotide sequences in double-stranded DNA and are a major tool of the biotechnologist. For prokaryotic cells, they function in nature as restriction-modification systems and will cleave foreign DNA that enters the bacterial cell (e.g., bacteriophage) but will not cleave host DNA that has been "protected" or modified by methylation. Most restriction endonucleases will reproducibly cleave DNA at a precise point within a recognition sequence. Generally, different enzymes will recognize different sequences that are four to six nucleotides long. This precision is essential for molecular cloning techniques, such as isolating genes from genomic DNA or inserting foreign DNA into a plasmid. Many enzymes make reproducible staggered cuts with respect to the dyad axis of symmetry of their recognition sequence. Cleavages of this type will yield DNA fragments with 3′ and 5′ overhangs or "sticky ends" that can base pair with DNA fragments generated by the same enzyme and thus form recombinant molecules.

Other enzymes will cut the recognition site *at* the axis of symmetry to yield blunt-ended cleavage products. These can be ligated to other blunt-ended fragments irrespective of the restriction enzyme used.

The second technique introduced in this exercise is agarose gel electrophoresis. When agarose is melted and then cooled in an aqueous solution, it forms a gel by hydrogen bonding. The population of DNA fragments generated by restriction enzymes will move through an agarose gel under the influence of an electric field, where negatively charged DNA molecules will be drawn to the anodes. Their rate of movement is based almost entirely on size, with the largest molecules having the lowest mobilities. The concentration of agarose in the gel determines pore size, and a DNA fragment having a particular size will migrate at different rates through gels of different concentrations.

There is a linear relationship between the log of mobility and gel concentration over a certain range of fragment sizes, so a gel concentration must be chosen that will effectively separate the molecules in the DNA population. Gels of 0.8% (w/v) agarose are suitable for separating linear DNA molecules 0.5–10 kb in size. The log molecular weights of the known marker fragments, such as λ phage cut with the restriction nuclease *Hin*dIII, can be plotted against mobility. The resulting calibration curve can be used to determine the molecular weights of unknown DNA fragments. As a quick measure, the molecular weight of an unknown fragment can be estimated by direct comparison by eye to the position of λ fragments on the gel. The bands are visualized by ultraviolet (UV) light illumination after staining with the fluorescent dye, ethidium bromide. (The appropriate quantity of DNA present in a band can be estimated as well by comparison of fluorescence intensities as in Exercise 5, Part C.)

Purity and any degradation of the DNA can also be determined on agarose gels. Contamination with RNA can be detected as a broad smear running at low molecular weight. RNA can be removed by addition of RNase to restriction digests. Contamination with protein may result in partial inhibition of restriction enzyme activity or degradation of the DNA. Protein contaminants can be removed by phenol–chloroform extraction and subsequent ethanol precipitation. Excess salt in a plasmid preparation will often inhibit enzymatic activity as well. Salt can be removed by ethanol precipitation followed by washing of a DNA pellet with 70% (v/v) ethanol. The agarose gel will show if the digested plasmid yields the bands expected from known restriction sites on the plasmid map. If the DNA has been degraded, e.g., by contaminating DNases, the DNA will appear as a smear in the stained gel.

In this exercise, we will use restriction enzymes and agarose mini-gels as a rapid diagnostic tool to determine the size and purity of both maxi-prep and mini-prep pRY121 DNA. Additionally, the approximate quantity of plasmid DNA in the samples can be determined by "band-brightness" comparison (see Exercise 5, Part C). The success of the PCR amplification of a *lacZ* gene fragment, as attempted in Exercise 7, will also be evaluated in this exercise.

## Reagents/Supplies

Agarose (SeaKem; FMC Corporation)
Bovine serum albumin (BSA; 10 mg/ml)
Dithiothreitol, 20× (DTT; 20 m$M$)
Erlenmeyer flasks (250 ml)
Ethidium bromide (10 mg/ml; see Exercise 5)
Gloves
Ice
Laboratory tape
λ *Hind*III size markers, concentration provided by the manufacturer (~0.5 μg/ml)
Microcentrifuge tubes (0.5 ml)
Micropippetor and tips
Pasteur pipet
PCR reaction samples (from Exercise 7)
Plastic wrap
Polaroid film (type 667)
pRY121 plasmid DNA from maxi-prep (from Exercise 6)
pRY121 plasmid DNA from mini-prep (from Exercise 4)

Restriction enzyme reaction buffers, 10× (see Instructor's Note, below)

Restriction enzymes: *Hin*dIII, *Pst*I, *Bam*HI, *Eco*RI (in glycerol; available from various manufacturers, see Appendix 20)

Stop buffer: 200 mM EDTA (pH 8.0)

TEA buffer, 25×: 1M Tris, 15 mM EDTA, 125 mM sodium acetate (pH 7.8). To 750 ml of distilled $H_2O$ add 121 g of Tris base, 10.2 g of sodium acetate, 18.6 g of EDTA. Adjust the pH to 7.8 with glacial acetic acid, and bring the volume to 1 liter. Store at 4°C

Tracking dye, 6× (see Instructor's Note, below)

Tupperware containers (or equivalent) for staining

## Equipment

Heat block at 65°C

Horizontal gel electrophoresis apparatus (see Instructor's Note, below)

Kodak 22A Wratten filter

Microcentrifuge

Microwave oven or Bunsen burner

Polaroid MP-4 camera

Power supply capable of delivering at least 100 volts

Ultraviolet (UV) transilluminator

Water bath at 37°C

Water bath at 45°C

### Instructor's Notes

1. Preparation of tracking dye solution: For 100 ml of a 6× stock solution, add 40 g of sucrose to 60 ml of distilled $H_2O$. Add 250 mg of bromphenol blue. Mix by stirring for 15–20 minutes. Store at 4°C. This 6× dye solution is added at one-sixth of the final volume of the DNA solution to be electrophoresed.

2. When purchasing a horizontal gel electrophoresis apparatus the instructor should consider the number of samples to be processed

in this exercise. Several manufacturers sell apparatuses sufficient for this purpose (see Appendix 20).

## Procedure

1. The following guidelines are used to prepare plasmid DNA for restriction digestion:

   a. Determine the quantity of DNA to be digested (i.e., 1 $\mu$g).

   b. On the basis of the stock concentration (units per microliter) of restriction endonuclease, calculate the microliters of enzyme required to achieve digestion of the desired quantity of DNA. For this experiment, 2 units of enzyme is sufficient to digest 1 $\mu$g of plasmid DNA. (*Note:* 6 units/$\mu$g is the "rule of thumb" for digestion of *chromosomal* DNA.) As use of excess enzyme ensures complete digestion, the manufacturer stock is often added as supplied and without dilution.

   c. Determine the total volume of the digestion reaction to attain an appropriate volume in which the glycerol content in the reaction will be less than or equal to 5% (v/v). Usually restriction nucleases are supplied in 50% (v/v) glycerol so that a 1:10 dilution of enzyme in the final reaction mixture would suffice.

   d. On the basis of this total reaction volume, calculate how much 10× buffer and 20× DTT (if required) is needed to achieve a 1× final concentration of these reagents in the reaction mixture. The buffer used is dependent on the salt concentration at which the enzyme is optimally active. The appropriate buffer (low, medium, or high salt) is supplied by the manufacturer.

   e. For some enzymes bovine serum albumin (BSA) is required as a stabilizer for optimal digestion. The BSA is often provided as a 100× stock solution at 10 mg/ml and should be added to reactions for a final concentration of 0.1 mg/ml. It is convenient to prepare a 10× working stock solution for this purpose. It is prudent to check whether an enzyme requires BSA or not prior to digestion. If BSA is added, a comparable quantity of distilled $H_2O$ should be withheld from the reaction mixture.

f. Calculate the volume of distilled $H_2O$ required to adjust the volume of the reaction to the desired final volume. See the sample tabulation (Table 8.1) for an example of how such a reaction may be prepared and properly recorded in a notebook.

**Table 8.1   Preparation of Reaction Mixture for DNA Digest by a Restriction Endonuclease**

| Tube | Concentration of DNA | DNA solution to give 1 μg of DNA | Enzyme and lot (concentration of enzyme) | Enzyme solution to give 2 U of DNA per microgram | 10× Buffer (salt concentration) | 10× BSA | Distilled water | Total volume |
|------|---------------------|----------------------------------|------------------------------------------|--------------------------------------------------|----------------------------------|---------|-----------------|--------------|
| 1 | 0.5 μg/μl | 2 μl | EcoRI, #61121 (10 U/μl) | 1 μl of a 1:5 dilution[a] | 2 μl (medium) | 2 μl | 13 μl | 20 μl |

[a] Although restriction enzymes may be diluted in 1× reaction buffer (*not water!*) to achieve an appropriate concentration, it is often advisable, for convenience, simply to add the enzyme in excess without diluting by using 0.5–1 μl of a stock of 10 U/μl in this case).

2. Calculate the amounts of the ingredients to be used in each of the following restriction digests: (1) pRY121 (mini-prep) cut with BamHI, (2) pRY121 (maxi-prep) cut with HindIII, (3) pRY121 (maxi-prep) cut with EcoRI, (4) pRY121 (maxi-prep) cut with PstI, and (5) pRY121 (maxi-prep) cut with BamHI.

### Instructor's Note

Manufacturers will supply a recommended buffer with the enzyme. Alternatively, these 10X buffers (referred to as low, medium, or high salt) can be prepared according to manufacturer specifications and stored in 1.5-ml aliquots at −20°C. A "universal buffer" suitable for most restriction endonucleases is also available from Stratagene. This buffer is useful for simultaneous digestion using multiple enzymes of different salt requirements. Likewise, BSA should or should not be added to the digest according to the enzyme supplier instructions for recommended buffer compositions.

3. Use a micropipettor to dispense ingredients into a 0.5-ml micro-centrifuge tube. Use a *fresh sterile tip* for each ingredient to avoid cross-contamination. The enzyme should be at −20°C until use and *should be added last*. All other ingredients should be kept on ice. The enzyme storage buffer is very viscous and will stick to the outside of the micropipet tip. To ensure accuracy, the very tip of the pipet should just touch the surface when drawing up the enzyme solution.

4. Mix the digest by gently flicking the microcentrifuge tube. Do not vortex.

5. Pellet any droplets by spinning the tube for 3 seconds in a micro-centrifuge at top speed.

6. Incubate the digests at 37°C in a water bath for 1 hour. During this hour, carry out Steps 7–11 as described below.

7. Assemble the minigel electrophoresis apparatus according to the manufacturer instructions.

8. Melt 0.8 g of agarose in 100 ml of 1× TEA buffer in a 250-ml flask and equilibrate to 45°C in a water bath. Melting is facilitated by heating in a microwave oven or by using a Bunsen burner.

9. Fill the trough or mold with agarose (45°C) and immediately insert the comb. Act quickly to remove any bubbles by touching them with a Pasteur pipet.

10. Allow the gel to solidify at room temperature for 20 minutes. It should be about 75 mm thick. Gently remove the comb.

11. Remove any tape from the trough and fill the buffer reservoirs with 1× TEA buffer until it just covers the gel.

12. Remove the reaction tubes from the water bath (see Step 6) and pellet the condensation in the tubes by centrifuging for 3 seconds at top speed in a microcentrifuge.

### Instructor's Note _____

The following step is optional. Add 1 $\mu$l of 200 mM EDTA (pH 8.0) stop buffer. The inactivation of restriction enzymes by chelation of metal cofactors prevents anomalous effects due to overexposure of DNA to active enzyme. For some restriction endonucleases other inactivating steps (i.e., heating at 65°C for 10 minutes) are recommended (see manufacturer suggestions).

13. Prepare the following solutions for electrophoresis.

   a. Prepare samples for analysis from the PCR reactions in Exercise 7 by removing 5 $\mu$l of sample from each reaction tube (pass the pipet tip through the mineral oil layer and into the aqueous layer) and pipetting the sample into a 0.5-ml microcentrifuge tube containing 15 $\mu$l of distilled H$_2$O. Mix by gently flicking the tube. [*Note:* If quantitating samples by band-brightness comparison (see Exercise 5), it may be desirable to prepare dilutions (1:2, 1:5, and 1:10 in 20 $\mu$l) of the PCR or restriction digest samples and run them on the gel as well, to provide for an accurate comparison.]

   b. Prepare a molecular weight standard for electrophoresis by adding 1 $\mu$l of λ *Hin*dIII-digested DNA (0.5-$\mu$g/ml stock) to 19 $\mu$l of distilled H$_2$O.

14. Add 4 $\mu$l of 6× tracking dye solution to all samples. This dye is also very viscous, and the micropipettor plunger should be depressed slowly to release the solution into the digest. Mix by gently pipetting.

15. Spin the tubes for 3 seconds at top speed in a microcentrifuge to rid the solution of bubbles.

16. Heat the samples at 65°C in a heat block for 2 minutes to dissociate DNA aggregates.

17. Use a micropipettor to load all the samples into the wells. It is convenient to correlate the tube number with the wells and order them from left to right. The orientation of your samples should be noted in your laboratory notebook. Work quickly and efficiently to avoid diffusion of the samples.

18. Connect the electrical leads [black is negative (−), red is positive (+)] so that the gel runs − to +. Negatively charged DNA molecules will migrate toward the + electrode (cathode). Some systems are manufactured with the leads linked to a cover that can attach to the gel box in only one way, thus making the hook-up foolproof.

19. Run the gel at 80 volts for about 1 hour, or until the bromphenol blue dye front is ~1.5 cm from the bottom.

## Safety Note

Do not touch the electrophoresis apparatus with the voltage on. Post *Warning High Voltage* signs.

20. Decant the buffer (reusable) and mark the −/+ orientation by cutting off a corner of the gel. Stain the gel in a Tupperware container for 10 minutes in ethidium bromide (0.5 μg/ml): add 5 μl of a 10-mg/ml stock to 100 ml of distilled water. Recall that ethidium bromide is *highly toxic. Wear gloves!*

21. Decant the ethidium bromide into an appropriate waste container (see Exercise 5 for inactivation and disposal). Add 100 ml of distilled water to the gel to destain. Agitate the gel gently for 5 minutes.

22. Using gloves, transfer the gel to a UV transilluminator set up beneath a Polaroid MP-4 camera and illuminate the bands under UV light. Load the film cassette with Polaroid type 667 film, according to manufacturer instructions.

23. Focus the camera on the gel. For convenience, position a ruler next to the gel and focus on the small print. It is important to

minimize exposure of both yourself and the gel to UV light. In practice, restriction fragments may be extracted from the gel for use in cloning experiments, and overexposure to UV light could cause undesirable nicks in the DNA.

24. A red filter (Kodak 22A Wratten filter) should be placed over the camera lens. An *f*-stop setting of 4.5 and an exposure time setting of B (1 second) gives a clear picture of the bands in the gel after development for 20 seconds.

25. Use the photograph to estimate the size fragments from each restriction digest by comparison to the λ *Hin*dIII size markers. Compare the fragments to those expected from the pRY121 restriction map (see Figure 8.1).

26. Refer to Exercise 5, Part C, for quantitation of DNA by "band-brightness" comparison and determine the approximate quantity

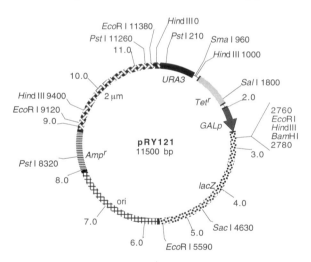

**Figure 8.1**   Map of plasmid pRY121. Location of the *Amp^r* (ampicillin resistance), *URA3* (production of uracil), *GALp* (galactose-inducible promoter), and *lacZ* (β-galactosidase) genes are indicated. The ori and 2 μm regions represent origins of replication for *E. coli* and yeast, respectively. *Tet^r* corresponds to a truncated gene that no longer confers tetracycline resistance. All cleavage sites for the specified restriction enzymes are also indicated. Consult West *et al.* (1984) for details on the construction of this plasmid.

of DNA in your plasmid preparations (both maxi-prep and mini-prep), using the photograph of your gel.

27. Make a note of the size of any PCR product(s) observed in the sample lanes containing aliquots from those reactions.

## References

Many manufacturer's catalogs, such as those from New England Biolabs, Life Technologies, Inc., Promega Corp., and Stratagene, are excellent sources of information on restriction enzymes and their uses. The manufacturers provide such catalogs on request for distribution to students.

West, R. W., Yocum, R. R., and Ptashne, M. (1984). *Saccharomyces cerevisiae GAL1–GAL10* divergent promoter region: Location and function of the upstream activating sequence UAS$_G$. *Mol. Cell. Biol.* **4**, 2467–2478. [This paper describes the construction and use of the plasmid vector pRY121.]

## Questions

1. Prepare a reaction mixture for a restriction digest containing components from the following stock solutions: DNA (2 mg/ml), BSA (10×), reaction buffer (10×), and *Bam*HI (10 U/$\mu$l). Indicate the minimum total reaction volume required for digesting 10 $\mu$g of plasmid DNA and the order in which these reagents should be added.

2. Calculate the expected sizes of the DNA fragments from the following digests of pRY121:

   a. *Sac*I

   b. *Hin*dIII and *Pst*I

   c. *Sma*I and *Eco*RI

   d. *Xho*I

3. What would be the effect on plasmid function of subcloning ("inserting") a 1.0-kb random gene fragment into:

   a. The *Sac*I site in pRY121?

   b. The *Sma*I site in pRY121?

   c. The *Pst*I site in the *URA3* gene in pRY121?

# Southern Transfer

## Introduction

Southern transfer enables the identification of a genetic sequence of interest within a complex mixture of DNA fragments. The DNA sample(s) is digested with restriction enzyme(s) to yield a set of DNA fragments that is then separated according to size by electrophoresis through an agarose gel. The goal of the Southern technique is to obtain a replica of the gel that retains the original positions of the DNA fragments. This is accomplished by transferring, or "blotting," the DNA from the gel onto a piece of transfer medium. Once immobilized on the transfer medium, the DNA can be probed for the sequence of interest.

The original protocol of Southern (1975) used a piece of nitrocellulose filter paper as the transfer medium. The more versatile nylon membrane material has become the preferred medium for many applications. For example, nylon is more amenable to "strip-washing," where the blot is cleared of the original probe by exposure to low-salt/high-temperature conditions for reprobing. Nylon also does not require baking to "fix" DNA onto the membrane, as does nitrocellulose. This cuts 2 hours off of the procedure time and avoids the problem of brittleness associated with baking. Several brands of membranes are sold by different companies [e.g., Nytran (Schleicher & Schuell), Immobilon (Millipore), and Zeta Probe (Bio-Rad)].

Blotting is accomplished as follows: Nicks are made in the DNA by exposing the gel to shortwave ultraviolet (UV) light for 20 seconds.

Next, the gel is soaked in alkali, which will denature the DNA by breaking hydrogen bonds. The combination of nicking and denaturation is important to facilitate transfer and provide single strands, which enables the DNA to hybridize with complementary single-stranded probe (see Exercise 10). The gel is then neutralized by soaking in Tris buffer, and is now ready to be "blotted."

The gel is positioned on top of the buffer-soaked filter and a piece of transfer medium is placed on top of the gel. This completes the gel "sandwich." Dry paper towels are then piled on top of the filter. Capillary action pulls the buffer from the bottom filter, through the gel, to the transfer medium and up through the paper towel stack. Rapid transfer methods by vacuum have been developed as well, but require the purchase of an additional apparatus. The DNA fragments are trapped on the transfer medium. The DNA is then permanently fixed to the nitrocellulose by baking, or to the nylon membrane by baking *or* cross-linking the strands with UV light. The replica can then be probed for the sequence of interest.

# Reagents/Supplies

## Part A.  Agarose Gel Electrophoresis

Agarose, molecular biology grade (SeaKem brand; FMC)
Bunsen burner (alternative to microwave oven)
DNA samples prepared by the instructor (see Instructor's Note, below)
Erlenmeyer flasks, 250 ml
Ethidium bromide (10 mg/ml)
Glass baking dishes (2 quart or equivalent)
Gloves
Micropipettor
Polaroid film, type 667
Spatula, large enough to handle gel
TEA buffer stock, 25× (see Exercise 8)
Tracking dye (see Exercise 8)

## Instructor's Notes

1. Gel apparatuses are available from many scientific suppliers at nominal prices. Any apparatus will suffice. However, we recommend a horizontal gel chamber, which enables pouring and running a 15 × 15 cm gel. If you wish to fabricate your own apparatus, refer to Sambrook *et al.* (1989).

2. Prepare samples of each of the following DNAs in a volume of 20 μl:

    a. λ phage DNA cut with *Hin*dIII—0.5 μg (available from various manufacturers)

    b. *Saccharomyces cerevisiae* whole-cell DNA cut with *Pst*I—1 μg (digest 1 μg of whole-cell DNA with 6 U *Pst*I; see Note 3 below)

    c. *lacZ* gene fragment amplified by PCR (from Exercise 7)— 5 μl of PCR mix, undigested (add 15 μl of distilled H₂O to this sample)

    d. Salmon testes DNA cut with *Pst*I—1 μg (see Note 2b above)

    e. Plasmid DNA (pRY121) cut with *Pst*I—1 μg [from student mini-prep (Exercise 6)]

3. *Saccharomyces cerevisiae* whole-cell DNA may be isolated by a simple procedure outlined in Kaiser *et al.* (1994). Alternatively, Gentra Systems, Inc., sells a rapid yeast genomic DNA isolation kit. Salmon testes DNA is available from Sigma.

## Part B. Southern Transfer

Aluminum foil

Denaturation buffer, 5 ×: 2 *M* NaOH, 4 *M* NaCl

Ethanol

Glass plates (as weights)

Glass rod or test tube

Gloves

Neutralization buffer, 1 ×: 0.5 *M* Tris (pH 7.6), 1.5 *M* NaCl

Nylon transfer membranes (0.2 μm) or nitrocellulose membrane (0.2 μm)

Paper towels (6 inches) cut to match the size of the agarose gel

Parafilm

Plastic boiling bags
Plastic wrap
Razor blade
SSC stock, 20 ×: 3 $M$ NaCl, 0.3 $M$ sodium citrate (pH 7.0)
Whatman 3 MM filter paper

## Equipment

### *Part A*

Gel electrophoresis apparatus (see Instructor's Note, above)
Kodak Wratten filter 22A
Microwave oven (alternative to Bunsen burner)
Polaroid MP-4 camera
Power supply (capable of delivering 50 volts)
Ultraviolet (UV) transilluminator
Water bath set at 45°C

### *Part B*

Heat sealer (e.g., Sears Seal-a-Meal) (optional)
Stratalinker cross-linking apparatus (Stratagene, Inc.) (optional)
Vacuum oven, 70°C (for nitrocellulose)
UV light source (hand held, short wave)

## *Procedure*

### Part A.  Agarose Gel Electrophoresis

Please follow the guidelines published by the manufacturer of the apparatus used at your laboratory.

### Day 1

1. Assemble the gel electrophoresis cell according to the directions of the manufacturer.

2. In a 250-ml flask prepare 100 ml of 1% (w/v) agarose in TEA buffer. [*Note:* 150–200 ml of 1% (w/v) agarose is usually sufficient for pouring a 15 × 15 cm gel.] Dissolve in a microwave oven at high power for 3–4 minutes. When the mixture begins to boil, turn off the oven immediately. Swirl and hold the flask to the light to inspect for refractile, undissolved agar particles. Return the flask to the microwave and heat until the molten agar is clear by inspection. Alternatively, agarose can be melted by swirling the flask over a Bunsen burner. Cool to a working temperature by placing the agarose in a 45°C water bath.

3. Add the 1% (w/v) agarose to the cell or gel mold by quickly pouring directly from the flask until the level reaches approximately 1 cm in thickness. The agarose should not be above 50°C, or it may shrink during solidification. Below this temperature, it may start to gel and must be remelted. Any bubbles that form will usually rise to the top.

4. Immediately insert a gel comb. *Avoid bubbles under the comb teeth.* The comb forms the sample wells and will cause some overflow. Allow the gel to solidify at room temperature for 30 minutes.

5. Attach the cell to, or place the mold in, the electrophoresis apparatus. Fill the reservoir with TEA buffer until the gel is immersed completely.

6. Remove the comb slowly and evenly by pulling it upward.

7. Add 6× tracking dye to the DNA samples provided by the instructor so that the final concentration of dye is 1×. Heat the samples for 2 minutes at 65°C.

8. Load the entire volume of the samples into the wells of the gel sample, using a micropipettor. Note the orientation of the samples. Work quickly and steadily to minimize sample diffusion, which occurs while the voltage is off.

9. Hook up electrical leads to the power supply (black = −, red = +) so that the samples run − to +. Run at 25 volts for 16 hours.

## Safety Note _____

Do not touch the electrophoresis apparatus with the voltage on. Post *Danger High Voltage* signs.

## Day 2

10. The bromphenol blue dye front should be ~2–2.5 cm away from the bottom edge of the agarose at 16 hours. Turn off the power, detach the leads, and decant the buffer. The buffer can be reused several times.

11. Cut off a small piece of gel from the upper right corner to mark the orientation.

12. Stain the gel by flooding it for 20 minutes in an 8 inch × 12 inch (2 quart) glass baking dish or equivalent containing 500 ml of distilled water and 25 $\mu$l of a 10-mg/ml stock solution of ethidium bromide. *Wear gloves!*

13. Using a spatula transfer the gel to another dish of distilled $H_2O$ for 15 minutes to destain. Destaining is facilitated by gentle agitation of the dish. Ethidium bromide waste should be disposed of as described in Exercise 5.

14. Slide the gel onto a transilluminator, using a spatula. Observe for banding patterns and photograph as described in Exercise 8.

## Part B.  Southern Transfer

## Day 2 (continued)

1. Expose the gel to UV light (hand held, shortwave) for 30 seconds to make nicks in the DNA. This facilitates transfer of DNA to nitrocellulose or nylon membranes.

2. Flood with 400 ml of 1× denaturation buffer. Shake gently for 30 minutes.

3. Decant the buffer and rinse the gel briefly in distilled water. Flood with 400 ml of 1× neutralization buffer. Shake gently for 20–30 minutes.

4. While waiting during Steps 2 and 3 (see Figure 9.1):

   a. Measure and cut a piece of transfer membrane that corresponds to the size of the gel [filters may be purchased precut or in sheets and rolls, which may be cut to size with a pair of scissors (ethanol sterilized) or a paper cutter]. (Handle all types of membranes with gloves.) If using nitrocellulose, wet it first in distilled water. Lower one edge, followed slowly by the rest so that it wets evenly. Once the nitrocellulose membrane is completely wet, transfer it to a container of 10× SSC for 5 minutes or until ready to use. Nylon membranes can be wet by placing them directly into a container of 10× SSC.

   b. Cut two pieces of Whatman 3 MM filter paper 1/8 inch larger than the membrane. Also cut a wick of Whatman 3 MM paper (see Figure 9.1).

   c. Cut Parafilm strips (1 inch × 6 inches), four per gel.

   d. Cut a 6-inch stack of paper towels to the size of the gel.

   e. In an 8 inch × 12 inch glass baking dish, place a piece of Plexiglass or stack of glass plates about 2 inches high. This platform should be about 1 inch longer and wider than the

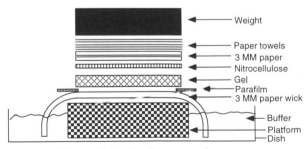

**Figure 9.1**   Southern transfer.

gel. Fill the dish with 10× SSC buffer to a level just below the top of the platform.

    f. Wet the Whatman 3 MM wick in 10× SSC and drape it over the platform so both ends are well submerged in buffer. (*Caution:* The wicks tear easily when wet.) Use a glass rod or test tube to roll over the surface to extrude bubbles.

5. Transfer the gel to the top of wick. Either side can face the membrane. Transfer the nylon (or nitrocellulose) to the top of gel. Avoid picking up and repositioning the gel because some transfer of DNA occurs immediately.

6. Wet two of the Whatman filters together in 10× SSC and position them on top of the membrane. Extrude bubbles from the Whatman filters as in Step 4f above.

7. Slip Parafilm strips on the wick along the edges of the gel. These will act as barriers to prevent short circuiting of the SSC buffer from the wick to the stack of towels if the towels inadvertently touch the buffer.

8. Stack 3–4 inches of paper towels on top of the filters.

9. Place an evenly distributed 1½-pound weight, such as several glass plates, on top of the towels.

10. Cover the whole dish containing the transfer assembly with plastic wrap to prevent evaporation. Allow the transfer to proceed overnight since fragments >10 Kb require ~15 hours to transfer completely.

**Note** _____

Capillary flow will cease should the *entire* stack of towels become wet. If necessary, cut more paper towels to size and replace wet towels with dry ones to continue the capillary flow.

## Day 3

11. Remove the paper towels and Parafilm strips.

12. Pick up the wick–gel–membrane–filter sandwich and invert. Remove the wicks. Mark the orientation on the membrane with a

lead pencil by noting the previously cut-off corner of the gel by well 1. Also mark the well positions. This is important, because in practice you may ultimately be measuring relative mobilities of DNA fragments from an autoradiogram or another detection system.

13. Peel off the gel. Because it contains EtBr, it should be treated as a biohazard and discarded accordingly. If desired, check for complete transfer of DNA by placing the gel on a transilluminator and look for ethidium bromide-stained DNA fragments that may not have transferred.

14. Because the outermost lanes are not flush with the edges of the gel, several millimeters of transfer medium may be considered excess, and can be cut off with an ethanol-sterilized razor blade. This step facilitates fitting the blot into plastic boiling bags (Step 16, below).

15. The DNA must be fixed to the nitrocellulose or nylon filters:

    a. For nitrocellulose filters, rinse the filters in 400 ml of 2× SSC to remove gel particles and blot excess fluid onto Whatman 3 MM paper. Place the filter between two pieces of Whatman 3 MM paper. Wrap in aluminum foil and bake in a 70°C vacuum oven for 1.5 hours. The heating fixes the DNA to the filter. The oven must be under vacuum or the nitrocellulose becomes brittle and cracks. The nitrocellulose is now stored until ready to probe.

    b. For nylon membranes fix the DNA by exposing the side that faced the gel to short-wave UV light for 20 seconds. Rinse in 2× SSC to remove gel particles. Allow the membrane to dry on a benchtop between two pieces of clean, dry Whatman 3 MM paper.

    c. Alternatively, an apparatus has been designed to facilitate the ease and reliability of UV cross-linking. The Stratalinker (Stratagene, Inc.) emits a uniform quantity of UV light and compensates for loss in UV light intensity over time to provide maximal cross-linking. It is effective for cross-linking of both nylon *and* nitrocellulose membranes.

16. The fixed membranes can now be sealed in household plastic boiling bags, using a heat sealer to protect them until ready to probe. The bags can be purchased presized or as a roll.

## References

Kaiser, C., Michaelis, S., and Mitchell, A. (1994). "Methods in Yeast Genetics." Cold Spring Harbor Laboratory Press, Cold Spring Harbor, New York, pp. 142–143. [This book provides a method for isolation of yeast genomic DNA for use in Southern blotting.]

Sambrook, J., Fritsch, E. F., and Maniatis, T. (1989). "Molecular Cloning: A Laboratory Manual," 2nd Ed. Cold Spring Harbor Laboratory, Cold Spring Harbor, New York, pp. 9.31–9.62. [Contains detailed protocols and discussions of DNA hybridization and blotting techniques.]

Southern, E. M. (1975). Detection of specific sequences among DNA fragments separated by gel electrophoresis. *J. Mol. Biol.* **98**, 503–512. [The classic paper that originally describes this method.]

## Questions

1. What is the purpose of UV-induced nicking and denaturation for facilitating transfer of DNA from agarose gels to membranes?

2. What are the applications of the Southern transfer procedure to molecular biology?

# *10*

# Preparation, Purification, and Hybridization of Probe

## Introduction

After DNA has been transferred onto nitrocellulose or nylon, detection of the DNA is carried out by hybridization to a probe DNA that is labeled radioactively or derivatized with a detectable group. Nick translation is a process whereby the probe DNA can be modified by incorporation of radioactive nucleotide or derivatized nucleotide. This technique will be employed to generate a labeled probe for hybridization to the blot prepared in Exercise 9.

For this exercise, a radioactive probe will not be used because of the regulations and potential hazards associated with radioactive phosporous, the usual nuclide used for incorporation into probes. If $^{32}$P labeling were used, autoradiography would follow hybridization to detect the transferred DNA–$^{32}$P-labeled probe hybrid. We will use a nonradioactive alternative that involves incorporation of biotinylated nucleotides into DNA by standard nick translation protocol, hybridization of the biotin-labeled DNA to the immobilized DNA, and detection of the hybrid by streptavidin and biotin-conjugated alkaline phosphatase (see Figure 10.1). In this detection system, the tetrameric streptavidin binds to the biotinylated probe in such a way that biotin-binding sites remain available to bind the biotin-modified alkaline phosphatase. Addition of a colorless substrate for alkaline phospha-

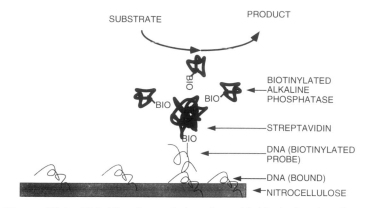

**Figure 10.1**   Hybridization and detection with biotinylated probe.

tase will indicate the site of the enzyme by production of a colored product. Thus, a "Dagwood sandwich" is formed *in situ* at the hybrid site that consists of the following: immobilized DNA–biotinylated DNA probe–streptavidin–biotinylated alkaline phosphatase–substrate/product.

In this exercise, we will prepare the biotinylated probe, purify the probe by standard protocols, and hybridize the probe to the Southern blot generated in Exercise 9. After hybridization, the method described above will be used to detect the DNA–DNA hybrid. Many of the reagents used are part of a commercially available kit from GIBCO/BRL (Life Technologies, Inc.). This system can detect 2–5 picograms (pg) of biotin-labeled probe. This level is sufficient for most probing experiments with a sensitivity that allows detection of a single-copy gene sequence in mammalian DNA. Other nonradioactive alternatives to DNA labeling and detection are available from various manufacturers (such as the Genius system from Boehringer Mannheim, the ECL sytsem from Amersham, the Phototype system from New England Biolabs, and the Lightsmith detection system from Promega). Some of these kits employ random primer labeling or end labeling as a means for obtaining labeled probes in conjunction with chemiluminescent detection methods. It is worthwhile considering alternate kits and methods prior to choosing a specific technique for labeling and detection, as they all have different advantages and disadvantages associated with them.

DNA may be conveniently labeled with biotin by nick translation in the presence of biotinylated adenosine triphosphate (biotin-14-dATP). The biotin-tagged nucleotide will be efficiently incorporated into DNA by DNA polymerase I in the presence of the other three unlabeled deoxynucleoside triphosphates. The sensitivity of detection of biotin-labeled DNA probes is not greatly affected by the degree of nucleotide incorporation beyond a value of 10–30 biotin-tagged nucleotides per kilobase of DNA.

Biotin-labeled DNA may be separated from unincorporated nucleotides by exclusion chromatography (ethanol precipitation may also be satisfactory). Gel filtration on Sephadex G-50 (coarse) is a rapid and quantitative procedure for recovery of labeled probe. Phenol extraction should always be avoided during handling and preparation of biotinylated probes because of possible partitioning of the probe into the phenol layer.

The purpose of passing the nick translation reaction over a gel-filtration chromatography column is to separate unincorporated nucleotides from the nick-translated probe. Spun columns of Sephadex G-50 offer a fast and easy method to achieve separation of molecules on the basis of size. Sephadex beads contain pores through which molecules under a certain hydrodynamic volume can pass. In this experiment, unincorporated nucleotides are small enough to pass in and out of the beads, whereas the probe is not. During the time of centrifugation (hence the name "spun"), the unincorporated nucleotides are held up in the column head matrix, while the probe can pass directly through and be eluted.

The hybridization kinetics of biotinylated DNA probes are virtually identical to those of radioisotope-labeled probes. The rate of solution hybridization is not significantly altered by incorporation of biotinylated nucleotides. However, standard hybridization conditions should be modified for use with biotin-labeled probes owing to a lower melting temperature ($T_m$) for the probe–target hybrid, as described by Leary *et al.* (1983). The steps outlined in Part C of this exercise use these modified protocols.

### Instructor's Note ─────────────────────────

An additional exercise using the biotinylated probe prepared in this exercise is found in Appendix 1, Exercise 10A, Colony Hybridization.

## Reagents/Supplies

### Part A.  Preparation of Probe

BioNick Labeling System (GIBCO/BRL)

Buffer 1: Tris-HCl (0.1 $M$, pH 7.5), NaCl (0.1 $M$), MgCl$_2$ (2 m$M$)

Buffer 2: Bovine serum albumin [3% (w/v); 3 g of BSA/100 ml of buffer 1]

Buffer 3: Tris-HCl (0.1 $M$, pH 9.5), NaCl (0.1 $M$), MgCl$_2$ (50 m$M$)

Column buffer: 50 m$M$ Tris (pH 7.9), 2 m$M$ EDTA (pH 8.0), 250 m$M$ NaCl

Components of the GIBCO/BRL BioNick Labeling System (purchased as a kit) that you will need for this protocol are as follows:

Control DNA: 5 $\mu$g of pBR322 plasmid DNA in 0.1 m$M$ EDTA, 10 m$M$ Tris-HCl (pH 8.0)

dNTP, 10 ×: 0.2 m$M$ concentration of dTTP, dGTP, and dCTP, 0.1 m$M$ dATP, 0.1 m$M$ biotin-14-dATP in 500 m$M$ Tris-HCl (pH 7.8), 50 m$M$ MgCl$_2$, 100 m$M$ 2-mercaptoethanol, and BSA (100 $\mu$g/ml)

Enzyme mix, 10 ×: DNA polymerase I (0.5 units/$\mu$l; BRL), DNase I (0.0075 units/$\mu$l), 50 m$M$ Tris (pH 7.5), 5 m$M$ magnesium acetate, 1 m$M$ 2-mercaptoethanol, 0.1 m$M$ phenylmethylsulfonyl fluoride (PMSF), 50% (v/v) glycerol, and nuclease-free BSA (100 $\mu$g/ml)

Stop buffer: 300 m$M$ EDTA (pH 8.0)

Ice

Micropipettor

SSC (20 ×): see Exercise 9

Triton X-100, 0.05% (v/v)

## Part B.  Purification of Probe

Buffers 1, 2, and 3 (see Reagents/Supplies, Part A)
Corex tubes (15 ml) and cushions
Glass wool, sterile
Sephadex G-50 (medium grade; Pharmacia) in column buffer
Syringes, 1 ml

## Part C.  Verification of Biotinylation Reaction

BluGene Non-Radioactive Nucleic Acid Detection System [including
   streptavidin, biotin (AP), NBT, BCIP; GIBCO/BRL]
Buffers 1, 2, and 3 (see Reagents/Supplies, Part A)
Pipet tip box tops
Plastic hybridization bags (GIBCO/BRL).
   *Note:* We recommend use of polypropylene bags, as high back-
   ground often results from use of other types (polyester)
Polypropylene tubes (12 × 75 mm)
Test strips of nitrocellulose or nylon

## Part D.  Hybridization of Probe

Biotinylated probe (from Part B of this exercise)
Denatured salmon sperm DNA, 100 $\mu$g/ml (Sigma)
Denhardt's solution, 50× (Ficoll, 5 g; polyvinylpyrrolidone, 5 g; BSA
   fraction V, 5 g; distilled H$_2$O to 500 ml); filter through a 0.45-$\mu$m
   membrane. Dispense in 25-ml portions and store at −20°C
Glass baking dishes (8 inches × 12 inches, 2 quart), or equivalent
Plastic hybridization bags (see Reagents/Supplies, Part C)
Plastic wrap
Prehybridization buffer, 1.5× [117 g of NaCl, 12.1 g of Tris base,
   0.7 g of EDTA, 2 g of sarkosyl (*N*-lauroylsarcosine), 2 g of sodium
   pyrophosphate to 1 liter, adjust pH to 7.9 with HCl, bring to 1.5
   liters with distilled water]

Southern blots of various DNAs (from Exercise 9)
SSC (2×) with SDS (0.1%, w/v)
Whatman 3 MM paper

# Equipment

## Parts A and B

Microcentrifuge
Tabletop centrifuge (Beckman TJ-6 or equivalent)
Water bath at 16°C (this can be set up with a thermostatted water
  bath in a cold room)

## Parts C and D

Heat sealer for plastic bags
Oven at 68°C
Rotary shaker at 68°C
Transfer plates (disposable incubation trays, Accutran; Sch-
  leicher & Schuell)
UV light (short wave)
Vacuum oven at 80°C
Water bath, boiling (100°C)
Water bath at 42°C
Water bath at 50°C
Water bath at 65°C

# Procedure

## Part A. Preparation of Probe

### Instructor's Note _____
Each group should set up a labeling reaction as described below with
both pRY121 DNA (from Exercise 6) and control pBR322 DNA.

## Day 1

1. Pipet the following reagents into a 1.5-ml microcentrifuge tube placed in ice:

   | | |
   |---|---|
   | 5.0 $\mu$l | 10× dNTP mix |
   | $x$   $\mu$l | Volume of control pBR322 DNA giving 1 $\mu$g of sample DNA |
   | $y$   $\mu$l | H$_2$O ($y$ is dependent on the volume of control or sample DNA) |
   | = 45   $\mu$l | Total volume |

2. Add 5 $\mu$l of 10× enzyme mix. Mix thoroughly, but gently, by flicking the tube. Centrifuge for 3 seconds at top speed in a microcentrifuge to bring liquid to the bottom of the tube.

3. Incubate at 16°C for 1 hour. During this time, complete Steps 1–12 in Part B (Purification of Probe) of this exercise.

4. Add 5 $\mu$l of stop buffer and mix by gently flicking the tube.

5. Add 50 $\mu$l of column buffer. Continue on to Part B, Step 13.

**Instructor's Note** _____
**Preparation of Sephadex G-50** Add 10 g of Sephadex G-50 beads (Pharmacia) to 250 ml of sterile water in a 500-ml bottle. Allow the beads to swell for 3 hours at room temperature. To hasten the swelling, the mixture may be heated to 65°C for 2 hours or to 90°C for 1 hour. Decant the supernatant (allow time for cooling if the latter methods are used) and add back an equal volume of column buffer. Repeat this wash two times and finally resuspend the slurry in an equal amount of column buffer. Sephadex G-50 columns, ready for use in DNA purification, are also commercially available.

## Part B. Purification of Probe

## Day 1 (continued)

**Safety Note** _____
Use gloves when handling glass wool.

1. Remove the plunger from a disposable 1-ml syringe.

2. Use the plunger (with rubber tip removed to avoid electrostatic attraction) to tamp a small ball of sterile glass wool to the bottom of the syringe.

3. Replace the rubber tip and retamp to ensure that the wool is adequately compressed.

4. Cut the cap off a microcentrifuge tube, put the tube into a 15-ml Corex tube, and set it in a rack.

5. Put the 1-ml syringe into the Corex tube so that the microcentrifuge tube will catch column effluent.

6. Add swollen Sephadex G-50 beads to the 1-ml mark on the syringe.

7. Balance the column apparatus.

8. Centrifuge at 1600 $g$ for 4 minutes in a tabletop centrifuge. The Sephadex will pack down.

9. Carefully remove the syringe and discard the contents of the microcentrifuge tube.

10. Reposition the microcentrifuge tube and syringe in the Corex tube.

11. Repeat Steps 6–10 until the level of packed Sephadex reaches the 0.9-ml mark.

12. Run through 100 $\mu$l of column buffer. Return to Part A, Step 4, of this exercise.

13. Cut the cap off a new microcentrifuge tube and set it up with the column in the Corex tube. Layer the nick translation reaction (in a volume of 100 $\mu$l) onto the column and centrifuge as in Step 8 above.

14. Transfer the effluent containing the nick-translated probe into an intact microcentrifuge tube. The purified probe is now ready for use. It can be stored at $-20°C$.

## Part C. Verification of Biotinylation Reaction

Day 2

1. To assure that biotinylated probe is obtained by this procedure, the following steps are taken (use biotinylated DNA for positive control and nonbiotinylated DNA as negative control).

2. Spot 2 $\mu$l of a 1:5 dilution series of prepared biotinylated pRY121, control biotinylated DNA, and control nonbiotinylated DNA onto three different test strips (approximately 10 cm × 0.5 cm) of nitrocellulose paper or nylon. Label each test strip and its orientation with a lead pencil and dry in a vacuum oven at 80°C for 30 min. Alternatively, to fix DNA to nylon, expose nylon to short-wave UV light for 20 seconds. *Do not touch filters with fingers. Wear gloves!*

3. Wash the test strips with buffer 1 for 1 minute. It is convenient to use transfer plates to ensure full immersion of strips in buffer.

4. For nitrocellulose blots, incubate in a water bath for 20 minutes at 42°C in buffer 2 (prewarmed to 42°C) and for nylon blots incubate at 65°C for 45 minutes in buffer 2 (prewarmed to 65°C).

5. Blot the test strips between two sheets of Whatman 3 MM and dry overnight at room temperature or in a vacuum oven at 80°C for 15 minutes. The dried strips can be stored desiccated for several months.

## Part C (continued). Verification of Probe

Day 3

6. Thoroughly rehydrate the test strips in buffer 2 for 10 minutes. Drain the buffer.

7. In a polypropylene tube dilute an appropriate volume of streptavidin to 2 $\mu$g/ml by adding 2 $\mu$l of stock solution (1 mg/ml) per 1.0 ml of buffer 1. Prepare approximately 1.0 ml per test strip.

Incubate the strips in diluted streptavidin for 10 minutes with gentle agitation, occasionally pipetting solution over the test strips. Decant the solution.

8. Wash the strips with buffer 1, using an at least 30-fold greater volume of buffer 1 than was employed in Step 7. A convenient container for this incubation is a pipet tip box top. Gently agitate the strips for 3 minutes in buffer 1. Decant the solution. Perform this wash step a total of three times.

9. In a polypropylene tube dilute an appropriate volume of biotin (AP) to 1 $\mu$g/ml by adding 1 $\mu$g of stock solution (1 mg/ml) per 1.0 ml of buffer 1. Prepare approximately 1.0 ml per test strip. Incubate the test strips in diluted biotin (AP) for 10 minutes, agitating gently, and occasionally pipetting solution over the strips. Decant the solution.

10. Wash the test strips with buffer 1, using an at least 30-fold greater volume of buffer 1 than was employed in Step 9. Gently agitate the test strips for 3 minutes in buffer 1. Decant the solution. Repeat this wash step once.

11. Perform Step 10 using buffer 3. Repeat this wash step once.

12. In a propropylene tube prepare approximately 7.5 ml of dye solution. The dye solution is prepared by adding 33 $\mu$l of NBT solution to 7.5 ml of buffer 3, gently mixing (by inverting the tube), and adding 25 $\mu$l of BCIP solution, followed by gentle mixing. *The dye solution should be prepared just prior to use.*

13. Incubate the test strips in the dye solution within a sealed polypropylene bag. Allow color development to proceed in the dark or in low light for several hours. Maximum color development usually is obtained within 4 hours. Nylon filters may be developed overnight.

14. Wash the test strips in 20 mM Tris (pH 7.5)–5 mM EDTA to terminate the color development reaction. Filters should be stored

dry and should always be protected from strong light. To dry, bake at 80°C in a vacuum oven for 1–2 minutes for nitrocellulose or for 2–5 minutes for nylon.

## Part D. Hybridization of Probe

### Day 4

1. Put the dry filter from the Southern blot (Exercise 9) into a plastic hybridization bag.

2. To the bag add 15 ml of 1.5× prehybridization buffer, 4 ml of 50× Denhardt's solution, and 0.2 ml of single-stranded, denatured salmon sperm DNA.

3. Squeeze the air out of the bag, and seal the end of the bag with a heat sealer. Incubate the bag by submerging it in a 68°C water bath for 2–4 hours (alternatively, a 68°C oven may be used if available). Occasionally agitate the bag to remove bubbles.

4. Add about 50 $\mu$l of purified probe to 400 $\mu$l of distilled water in a microcentrifuge tube. Place the tube in a boiling water bath for 2–4 minutes. Cut open a corner of the bag, add the probe, and reseal it with the heating unit.

5. Incubate the bag in a rotary oven at 68°C overnight.

### Day 5

6. Remove the filter from the bag and put it in a glass baking dish (2 quart) with 250 ml of 2× SSC with 0.1% (w/v) SDS. Incubate for 3 minutes at room temperature. Repeat this wash. Put the filter in 250 ml of 0.2× with 0.1% (w/v) SDS for 3 minutes at room temperature. Repeat this wash. Put the filter in 250 ml of 0.16× SSC with 0.1% (w/v) SDS for 15 minutes at 50°C. Repeat this wash. Briefly rinse the filter in 2× SSC with 0.1% (w/v) SDS at room temperature.

7. Dry the filter by blotting on Whatman 3 MM paper (do not dry out the filter).

8. Wrap the filter in plastic wrap and store in your desk.

9. To detect hybridized probe, carry out Steps 4 and 7–14 outlined in Part C of this Exercise (Verification of Probe) and adjust the volumes of detection solutions to 3.0 ml/100-cm$^2$ filter.

## References

Kricka, L. J. (1992). "Nonisotopic DNA Probe Techniques." Academic Press, San Diego, California. [This text is a very useful general reference describing various methods for nonradioactive labeling and detection of DNA.]

Langer, P. R., Waldrop, A. A., and Ward, D. C. (1981). Enzymatic synthesis of biotin labelled polynucleotides: Novel nucleic acid affinity probes. *Proc. Natl. Acad. Sci. U.S.A.* **78,** 6633–6637. [This paper describes the synthesis of biotinylated nucleotides for their use in tagging nucleic acids with biotin.]

Leary, J. J., Brigati, D. J., and Ward, D. C. (1983). Rapid and sensitive colorimetric method for visualizing biotin labelled DNA probes hybridized to DNA or RNA immobilized on nitrocellulose: Bioblots. *Proc. Natl. Acad. Sci. U.S.A.* **80,** 4045–4049. [Describes the use of biotinylated DNA as nonradioactive probes for nucleic acid hybridizations.]

## Questions

1. What are the components of the nick translation reaction? How do they function together to generate a probe?

2. What are the components of the biotinylated DNA detection system and what is their function?

3. How would you change the components of the hybridization reaction to increase or decrease the amount of hybridization?

# 11

# Transformation of *Saccharomyces cerevisiae*

## Introduction

*Saccharomyces cerevisiae* is a valuable organism to the field of biotechnology; it is a eukaryote, yet it can be cultivated like a prokaryote owing to its microbial characteristics. Because *S. cerevisiae* has a small genome, relatively short doubling time, and can be analyzed genetically, many of the advances that have been made in molecular biology have used this yeast as a research tool. This organism has important industrial uses as well for production of beer, bread, wine, and recombinant peptides and proteins.

The use of *S. cerevisiae* as a host organism for the expression of foreign DNA, introduced in the form of plasmid DNA vectors, has been an important part of current advances that have taken place in recombinant DNA technology (Botstein and Fink, 1988). Transformation techniques are similar to those applied for *Escherichia coli,* but are modified to account for differences in the cell wall complexity of yeast. Transformation of yeast is usually performed in one of two ways: either by the formation of spheroplasts, including the removal of the cell wall (Hinnen *et al.,* 1978), or by a more rapid treatment of intact cells with alkali cations (Ito *et al.,* 1983). Spheroplast formation has inherent difficulties associated with it that are related to the osmotic stability of the cells and tedious procedures.

Electroporation techniques have been utilized for transformation of yeast as well (Becker and Guarente, 1991) and offer the advantage of using smaller quantities of DNA to achieve transformation, but often exhibit strong strain-specific preferences in their effectiveness and require the use of an expensive apparatus. Improvements to the alkali cation method (Gietz *et al.,* 1992) that render it simpler and more effective have made it the method of choice for researchers in the field.

By exposing intact yeast cells, which have been harvested during the logarithmic phase of growth, to conditions of alkali cations (i.e., lithium acetate, rubidium chloride), heat shock, and polyalcohol treatment, changes can be induced in the cell envelope that facilitate plasmid uptake. The addition of carrier DNA also promotes the uptake of vector DNA. Carrier DNA allows complexing of small DNA molecules with larger carrier DNA molecules. The resulting larger complex reaches the cell surface more easily for subsequent uptake. As with *E. coli,* various factors influence transformation efficiency (measured as transformants per microgram of DNA; also see Exercise 12, Introduction). These include plasmid size, DNA configuration and quality, host strain, and selection procedures. Nevertheless, transformation of *S. cerevisiae* is attainable with efficiencies ranging between $10^4$ and $10^5$ transformants per microgram of DNA for some plasmids and strains.

The pRY121 plasmid used in this course serves as a useful instructional tool. It is a "shuttle vector" (replicates in two organisms, in this case *E. coli* and *S. cerevisiae*) and encodes an expressible protein, $\beta$-galactosidase, which is detectable by facile assays. Furthermore, the expression of this protein is coupled to a controlling genetic element, the *GAL* promoter. A description of the plasmid is given in Exercise 8.

## Reagents/Supplies

Centrifuge tubes, disposable, sterile, 50 ml (Falcon, type 2098)
Hemocytometer (see Appendix 14, Determination of Cell Numbers)

Micropipettor

Parafilm

Polyethylene glycol (PEG), 50% (w/v), autoclaved.

[*Note:* It is critical to use PEG with an average molecular weight of 3350 (Sigma), as use of higher molecular weight PEG decreases transformation efficiencies. It has also been noted that different batches of PEG can dramatically affect successful transformation and, thereby, represents a source of potential problems]

pRY121 plasmid DNA, purified by column chromatography or CsCl (from Exercise 6 or Exercise 6A in Appendix 1). (*Note:* A plasmid concentration of 1–10 $\mu$g/ml is required for optimal transformation of yeast. If the available samples of pRY121 are too dilute, they should be precipitated using ethanol and resuspended to 1–10 ng/ml)

*Saccharomyces cerevisiae* strain YNN281 (see Exercise 3 and Appendix 6)

Single-stranded, denatured salmon sperm DNA (Sigma), 10 mg/ml

Sterile water

TE buffer, pH 8.0; sterilize by autoclaving (see Exercise 4)

TE/LiAc [10 m$M$ Tris-Cl, 1 m$M$ EDTA, 100 m$M$ lithium acetate (pH 7.5)]; filter sterilize through a 2.0-$\mu$m filter

YEPD broth (see Exercise 2)

YNB minimal agar plates, selective for transformants (see Step 13 below and Exercise 2)

## Equipment

Incubator at 30°C

Microcentrifuge

Tabletop centrifuge or equivalent

Water bath at 42°C

Water bath or incubator shaker at 30°C

### Instructor's Notes _____

1. The day before this exercise is to be performed, aseptically inocu-

late 5 ml of YEPD broth in a sterile test tube with a colony of *S. cerevisiae* YNN281 cells, preferably from a fresh streak plate (1–2 weeks old). Prepare one 5-ml culture for each group of students. Incubate by shaking the tube overnight at 200 rpm in a water bath or incubator shaker set at 30°C. After approximately 12 hours of incubation, cells should be at or near a density of $1 \times 10^8$ cells/ml by class time.

2. Approximately 4–5 hours before class, prepare one subculture of this overnight culture for each group by inoculating 100 ml of YEPD broth in a 500-ml flask at a density of $2 \times 10^6$ cells/ml. Shake this flask in a water bath set at 200 rpm at 30°C. Assuming a doubling time of 90–100 minutes in YEPD, these cells should reach a density of approximately $1$–$2 \times 10^7$ cells/ml by class time.

3. To facilitate absorption of cells into the agar at the final plating step of this procedure, it helps to place all of the agar plates to be used in an incubator for 12–18 hours to allow them to dry slightly.

## Procedure

1. After the *S. cerevisiae* cells have reached an appropriate density ($1$–$2 \times 10^7$ cells/ml), aseptically split the culture into two sterile 50-ml disposable centrifuge tubes. Balance these tubes and harvest the cells by centrifugation in a tabletop centrifuge at 2000 rpm for 5 minutes at room temperature. Unlike bacteria, yeast are larger and, therefore, do not require higher speeds to pellet cells.

2. Discard the culture supernatant from each tube. Wash the cells to remove all traces of the rich YEPD broth by resuspending each cell pellet in 25–50 ml of sterile water and centrifuging this suspension at 2000 rpm for 5 minutes at room temperature.

3. Discard the supernatants. Add 1 ml of sterile water to one of the cell pellets and resuspend the cells by vortexing or repeated pipetting. Transfer this suspension to the remaining cell pellet in

the other centrifuge tube and resuspend the combined cells. Using a micropipettor, transfer the combined suspension to a sterile 1.5-ml microcentrifuge tube.

4. Pellet the yeast cells by centrifugation in a microcentrifuge for 20 seconds at top speed at room temperature. Decant or aseptically aspirate off the aqueous supernatant and add 1 ml of TE/LiAc to the cells. Resuspend the cell pellet by repeated pipetting and determine the concentration of cells by counting with a hemocytometer (see Appendix 14).

   *Note:* It can also be assumed that the cells present at this stage represent 100 ml of cells at an original density of $1-2 \times 10^7$, only concentrated into a volume of approximately 1 ml. Therefore, the total cell number should be close to $1-2 \times 10^9$.

5. Pellet the cells by centrifugation in a microcentrifuge for 20 seconds at top speed at room temperature. Decant or aspirate off the supernatant. Using the value derived from the cell count in Step 4, resuspend the cells at a density of $2 \times 10^9$ cells/ml in TE/LiAc. *It is best to treat the cells gently while resuspending them, as they are fragile at this stage.*

6. For each sample to be transformed, pipet 100 $\mu$l of cell suspension into a sterile 1.5-ml microcentrifuge tube. Your experiment should include preparation of enough tubes for transformation of at least one sample of pRY121 DNA and a control.

7. To each sample of cells, add 50 $\mu$g of single-stranded sperm carrier DNA. Also add 1-10 $\mu$g of pRY121 plasmid DNA in a maximum volume of 10 $\mu$l. Several samples covering a range of different pRY121 DNA concentrations between 1 and 10 $\mu$g is suggested. Add 10 $\mu$l of sterile water and 50 $\mu$g of carrier DNA to the control cells (no plasmid).

8. Add 240 $\mu$l of 50% (w/v) PEG and 60 $\mu$l of TE/LiAc to each sample and mix thoroughly by gentle and repeated pipetting. Incubate the samples at 30°C in a water bath for 30 minutes while shaking at 200 rpm.

9. Heat shock the samples by placing them in a 42°C water bath (stationary) for exactly 15 minutes. *Increased or decreased heat shock times reduce transformation efficiency.*

10. Pellet the cells by centrifugation in a microcentrifuge for 5 seconds at top speed at room temperature. Remove the supernatant solution by aspiration and resuspend the cells with 1 ml of TE buffer (pH 8.0) by gentle and repeated pipetting.

11. Pellet the cells again by centrifugation in a microcentrifuge for 5 seconds at top speed at room temperature. Remove the supernatant solution by aspiration and resuspend the cells with 1 ml of YEPD broth by gentle and repeated pipetting. Incubate the samples at 30°C for 45 minutes in a water bath while shaking at 200 rpm.

12. Wash the rich YEPD broth from the cells by repeating Step 10.

13. Plate 100 and 300 $\mu$l of the cells onto separate minimal agar plates that are selective for transformants (i.e., YNB agar without amino acids, but containing all the appropriate supplements for growth of strain YNN281 except for uracil). In addition, plate out the control untransformed cells (to which no plasmid DNA was added) onto the same selective minimal agar plates (negative control for growth) and also onto plates that contain uracil (postitive control for growth).

14. Allow the plates to dry and place them in an incubator, in an inverted position, at 30°C for 3–5 days. Transformants will appear as distinct colonies, whereas the background of nongrowing yeast cells appears as a hazy smear on the agar. Wrap these plates with Parafilm and store at 4°C.

## Instructor's Note

At this point in the course, it is strongly advised that the instructor or students prepare a streak plate of *S. cerevisiae* YNN281 (pRY121) transformants for use in subsequent exercises. Approximately four or five transformants should be streaked onto minimal medium selective for plasmid maintenance. Plates should be grown for 2–3 days at 30°C and then stored at 4°C, wrapped in Parafilm.

# References

Becker, D. M., and Guarente, L. (1991). High-efficiency transformation of yeast by electroporation. *Methods Enzymol.* **194,** 182–187. [This represents an alternative procedure for yeast transformation, which involves a technique of electric shock to facilitate DNA entry termed *electroporation.*]

Botstein, D., and Fink, G. R. (1988). Yeast: An experimental organism for modern biology. *Science* **240,** 1439–1443. [This article is an excellent review on why researchers study yeast and how this organism can be utilized to answer important biological questions.]

Gietz, D., St. Jean, A., Woods, R. A., and Schiestl, R. H. (1992). Improved method for high efficiency transformation of intact yeast cells. *Nucleic Acids Res.* **20,** 1425. [This reference is a widely used procedure for yeast transformation. This exercise was based on the methods described in this paper.]

Hinnen, A., Hicks, J. B., and Fink, G. R. (1978). Transformation of yeast. *Proc. Natl. Acad. Sci. U.S.A.* **75,** 1929–1933. [Paper describing the original method of yeast transformation that involves the enzymatic removal of the yeast cell wall to facilitate the DNA entry.]

Ito, H., Fukuda, Y., Kousaku, M., and Kimura, A. (1983). Transformation of intact yeast cells treated with alkali cations. *J. Bacteriol.* **153,** 163–168. [The paper describing transformation of whole, intact yeast cells.]

# Questions

1. List and describe four factors influencing transformation efficiency in yeast.

2. Calculate the transformation efficiency achieved in this exercise. Compare this value to that of other students in your class and propose reasons for any differences.

# 12

# Isolation of Plasmid from Yeast and *Escherichia coli* Transformation

## Introduction

While the yeast *Saccharomyces cerevisiae* is an excellent host for the expression and analysis of genes, it does have associated with it some technical disadvantages. Among these are its propensity for recombination, which can lead to gene rearrangements in plasmid constructs designed for heterologous expression. Additionally, analysis of foreign gene expression and function or phenotypic complementation of specific mutations in yeast often requires verification that a particular phenotype is plasmid-borne. Unfortunately, unlike *Escherichia coli,* in which the majority of chromosomal DNA is easily separated from plasmid DNA by virtue of its association with the bacterial cell membrane, pure yeast plasmid mini-preps are more difficult to prepare owing to contamination with mitochondrial and genomic DNA. Therefore, to confirm plasmid-related phenotypes or authenticate plasmid integrity, investigators usually shuttle plasmid DNA to *E. coli* cells to successfully recover and subsequently analyze any genes of interest in a given vector.

This exercise demonstrates a rapid procedure for liberating small quantities of plasmid DNA from yeast cells and immediately using this crude preparation for transformation of *E. coli.* Yeast cells are physically disrupted, lysed, and plasmid DNA is then solubilized by solvent extraction of the resulting mixture in the presence of LiCl and a nonionic detergent (Triton X-100). The DNA is ethanol precipitated and used directly to transform *E. coli.*

The most common method of bacterial transformation utilizes high levels of $CaCl_2$ to facilitate the entry of plasmid DNA vectors into *E. coli* in conjunction with structural alteration of the bacterial cell wall. In general, there are three basic steps in the introduction of plasmid DNA into cells: (1) preparation of "competent" cells (cells able to accept DNA), (2) transformation of these competent cells, and (3) selection of transformants. Most *E. coli* strains exhibit transformation efficiencies (measured as transformants per microgram plasmid DNA) in the range of $10^5$ to $10^8$. The factors influencing this efficiency are often related to conditions that render cells competent. These include using cells that are harvested during the logarithmic phase of growth, maintaining cells at a temperature of 4°C during treatment, and prolonged exposure to ice-cold $CaCl_2$. Following the preparation of competent cells, transformation is induced by a destabilization of the lipids in the cell envelope through heat shock treatment. A subsequent period of recovery time is then allotted to enable cells to begin expression of a selectable marker, usually antibiotic resistance. This period is followed by a plating of cells onto selective medium that will only allow survival of sucessfully transformed hosts. Transformation efficiencies differ between strains and vectors as well. Factors such as plasmid size, quantity, purity, DNA configuration (linear, closed circular, nicked), and selectable marker all influence the outcome of a transformation experiment. As this experiment is designed only for the recovery of a crude plasmid DNA preparation from yeast into *E. coli,* the transformation efficiency is very low, as only a few clones are usually obtained.

The final part of this exercise demonstrates the use of an alkaline lysis procedure for mini-prep isolation of plasmid DNA from the *E. coli* transformants. This is included as an alternative to the "boiling" mini-prep procedure previously utilized in Exercise 4. Isolation of mini-prep DNA can be coupled with agarose gel electrophoresis and/

or restriction enzyme analysis (as in Exercise 8) to verify the presence of plasmid DNA in the original yeast transformants. In all, while plasmid shuttling from yeast to *E. coli* is an inefficient procedure with regard to transformation, it is representative of the interchangeable nature of the host organisms and their respective utility in molecular genetic analyses.

# PART A    PREPARATION OF COMPETENT CELLS OF *Escherichia coli*

## Reagents/Supplies

Calcium chloride ($CaCl_2$), 100 m*M*, ice cold
Centrifuge tubes, 50 ml, Oak Ridge round bottom, polycarbonate (Nalgene), sterilized. (*Note:* Although these tubes are sold and designated as 50 ml, their actual capacity is ~35 ml)
*Escherichia coli* LE392 (see Exercise 1 and Appendix 7)
Ice
LB broth (see Exercise 1)
Test tube, 5 ml, sterile

## Equipment

Beckman JA-20 centrifuge rotor or equivalent
Beckman J2-21 centrifuge or equivalent
Water bath or incubator shaker, set at 37°C

## Procedure

### *Instructor's Notes* _____

1. The day before this exercise is to be performed, inoculate a 5-ml test tube containing LB broth with a colony of *E. coli* LE392

(preferably from a streak plate no more than 2 weeks old). Incubate at 37°C overnight (about 16 hours) in either a water bath or incubator, shaking at about 200 rpm.

2. The day of class, read the $OD_{600}$ of the overnight culture (a $1:10$ dilution usually suffices). Add enough cells to 40 ml of LB broth in a 500-ml flask to give an initial absorbance reading of approximately $OD_{600} = 0.1$. Incubate this flask with shaking at 37°C for 2–3 hours until class time.

## Day 1

1. Check the $OD_{600}$ of the culture against a blank of LB broth (this may require a $1:2$ or $1:3$ dilution for an accurate reading). When the cell density reaches $OD_{600} = 0.7$–$0.8$, harvest the culture by pouring the cells into an Oak Ridge polycarbonate centrifuge tube (as much as will fit into the tube) and centrifuge at 4°C for 5 minutes at 5000 rpm in a Beckman JA-20 rotor. Carefully pour off and discard the supernatant.

2. Resuspend the cells in 1 ml of ice-cold 100 m$M$ $CaCl_2$ by gently swirling the tube repeatedly. *Keep the tube cold by returning it to an ice bucket every 20–30 seconds.* Resuspension may also be facilitated by gently pipetting up and down with a prechilled pipet. After cells are completely resuspended, add an additional 19 ml of ice-cold 100 m$M$ $CaCl_2$, mix by gentle inversion, and place on ice for 20 minutes. $CaCl_2$ "coats" the cells with an overall positive charge that serves to attract the negatively charged DNA.

3. Centrifuge the cells again at 5000 rpm for 5 minutes at 4°C. Discard the supernatant and gently resuspend the cells in 1 ml of 100 m$M$ $CaCl_2$. Add an additional 3 ml of ice-cold 100 m$M$ $CaCl_2$ and mix by gently swirling. Cells should be placed on ice until needed for Part C of this exercise. The cells are now considered competent for DNA transformation. While these cells can be used immediately, for most *E. coli* strains, competency peaks at about 16–18 hours and begins to drop off significantly beyond 24 hours in $CaCl_2$.

## PART B     ISOLATION OF YEAST DNA AND TRANSFORMATION OF *Escherichia coli*

### Reagents/Supplies

Chloroform [24:1 (v/v), chloroform:isoamyl alcohol]
Competent *E. coli* LE392 cells from Part A of this exercise
Ethanol, 100%, ice-cold
Glass beads, 0.45–0.50 mm in size (B. Braun Biotech)
Gloves
LB agar plates containing ampicillin (50 $\mu$g/ml)
LB agar plates containing no antibiotic
Lysis buffer: 2.5 *M* LiCl, 50 m*M* Tris-HCl (pH 8.0), 4% (v/v) Triton X-100, 63 m*M* EDTA
Microcentrifuge tubes, 1.5 ml
Micropipettor
Parafilm
Phenol
Plasmid pRY121 maxi-prep DNA, concentration of at least 50 $\mu$g/ml (from Exercise 6)
*Saccharomyces cerevisiae* YNN281 transformed with pRY121 (from Exercise 11)
Spreader, hockey stick style
TE buffer, pH 8.0 (see Exercise 4)

### Equipment

Microcentrifuge
Vacuum centrifuge or vacuum desiccator
Water bath or incubator shaker, set at 30°C
Water bath, set at 37°C
Water bath, set at 42°C

### *Instructor's Notes* _____

1. About 16–18 hours before class, inoculate 3 ml of minimal me-

dium [YNB (without amino acids, with ammonium sulfate) containing tryptophan, histidine, lysine, and adenine; be sure to *omit* uracil: see Exercise 2] with a colony of *S. cerevisiae* YNN281 transformed with pRY121. The source of this inoculum should be the yeast transformants obtained in Exercise 11; however, it is best if several of these transformants (two to four per group) are streaked onto a fresh minimal agar plate 3–5 days before this exercise is to be performed to provide a fresh source of YNN281 (pRY121) and confirm the putative identity of the transformants. Grow this culture at 30°C while shaking at 200 rpm.

2. This procedure has been successfully performed using *several* colony transformants of yeast directly as a source of DNA, instead of an overnight culture. This method can be thought of as an option, if necessary, but is not considered optimal.

3. It is best to prepare the agar plates used for transformation at least a day before the start of the laboratory session and allow them to dry slightly by incubation on a benchtop or in an incubator (30–37°C) for about 16–24 hours.

## Procedure

### Day 1

1. Fill two 1.5-ml microcentrifuge tubes to the top with cells from an overnight culture of *S. cerevisiae* YNN281 (pRY121). Spin these tubes at top speed at room temperature in a microcentrifuge for 20 seconds. Aspirate and discard the supernatant.

2. After removing the supernatant, resuspend each cell pellet in 100 $\mu$l of lysis buffer by gently pipetting the mixture up and down repeatedly. Try to avoid forming bubbles.

3. Add an equal volume of phenol–chloroform [1 : 1 (v/v) mixture] and then 0.2 g of glass beads to both mixtures. *Do not use excess beads, as this will prevent adequate mixing of the solution described below.*

4. Vigorously vortex each mixture for 3 minutes to physically disrupt the integrity of the yeast cells.

## Safety Note _____

Wear gloves, as phenol and chloroform are toxic and can leak out of the tubes at times.

5. Centrifuge the tubes for 1 minute at top speed at room temperature in a microcentrifuge. Transfer the upper aqueous phase of each tube and combine into one fresh 1.5-ml microcentrifuge tube.

6. Add 500 $\mu$l of ice-cold ethanol. Vortex briefly to mix the contents and centrifuge at top speed at room temperature in a microcentrifuge for 5 minutes.

7. Aspirate off the ethanol supernatant and dry the nucleic acid pellet in a vacuum centrifuge or vacuum desiccator for about 5 minutes. Alternatively, invert the tube on absorbent paper and allow it to air dry for 15 minutes.

8. Resuspend the pellet in 30 $\mu$l of TE buffer. Place this tube on ice.

## Instructor's Note _____

As an interesting aside, it is also possible to utilize the yeast DNA isolated in the steps above as a source of template for PCR. If desired, a useful exercise is to have the class perform PCR, as outlined in Exercise 7, using various quantities of yeast DNA to generate a *lacZ* gene fragment as a product. We have found that dilution of the DNA (1:10–1:100) often facilitates successful PCR, as these DNA preparations are crude and appear to contain substances that inhibit *Taq* polymerase activity. Reaction products may be subsequently analyzed by agarose gel electrophoresis, as outlined in Exercise 8. This represents an alternative means of verification of plasmid DNA that can be substituted for the mini-prep procedure described in Part C of this exercise.

9. Pipet gently 300 μl of competent cells of *E. coli* (from Part A above) into each of six 1.5-ml microcentrifuge tubes labeled A, B, C, D, E, and F.

10. Add the following quantity of the yeast DNA preparation to each tube:

    Tube A:   20 μl
    Tube B:   5 μl
    Tube C:   1 μl
    Tube D:   1 μl of a 1 : 5 dilution (1 μl of yeast DNA + 4 μl of sterile distilled H₂O)
    Tube E:   1 μl of pRY121 maxi-prep DNA (from Exercise 6)
    Tube F:   No DNA (control)

11. Place the cells on ice for 30 minutes.

12. Heat shock the cells by placing them in a water bath at 42°C for 2 minutes. *Do not heat longer than this, as the competent cells are very sensitive to heat and will die.* Return the cells to ice for 4 minutes immediately after heat shocking them.

13. Add 700 μl of LB broth to each tube and mix by gently inverting the tubes three or four times. Incubate the tubes for 45 minutes at 37°C to allow for expression of ampicillin resistance from the plasmid.

14. Label five LB plates containing ampicillin alphabetically (A through E), and label one LB plate (without ampicillin) as the control (F). Spread 200 μl from each reaction tube (A–E) onto a correspondingly labeled LB ampicillin plate. Additionally, spread 200 μl of the control transformation F onto the control LB plate (F).

15. Allow the plates to dry on the benchtop until most of the liquid appears to have been absorbed into the agar, then incubate the plates at 37°C for 14–20 hours. Typically, only a few transformants result from plasmid shuttling. Plates should then be wrapped in Parafilm and stored at 4°C until they are used in Part C of this exercise.

## PART C    ALKALINE LYSIS MINI-PREP OF *Escherichia coli* DNA

### Reagents/Supplies

Chloroform (optional)
Culture tubes, 5 ml, disposable, sterile
*Escherichia coli* LE392 (pRY121) transformants from Part B of this exercise
Ethanol, 70% (v/v), ice-cold
Isopropanol (2-propanol), 100%
LB broth containing ampicillin (50 $\mu$g/ml)
LiCl, 5 $M$
Microcentrifuge tubes, 1.5 ml
Micropipettor
Paper towels
RNase A, 10-mg/ml stock (see Exercise 4)
Solution I: 50 m$M$ glucose, 25 mM Tris-Cl (pH 8.0), 10 m$M$ EDTA (pH 8.0)
Solution II: 0.2 N NaOH, 1% (w/v) SDS; prepare fresh the day this exercise is to be performed
Solution III: 60 ml of Potassium acetate (5 $M$), 11.5 ml of glacial acetic acid, 28.5 ml of distilled H$_2$O
TE buffer, pH 8.0 (see Exercise 4)

### Equipment

Beckman TJ-6 tabletop centrifuge or equivalent
Heat block, set at 55°C
Microcentrifuge
Water bath or incubator shaker, set at 37°C

### Procedure

#### Instructor's Notes _____

1. For each group of students, randomly pick four *E. coli* LE392

(pRY121) transformants from the LB ampicillin plates from Part B of this exercise and use them to inoculate four sterile disposable tubes containing 2.5 ml of LB broth. Shake these cultures at 37°C overnight until the start of class. Do not allow cultures to overgrow (usually 12–14 hours is sufficient), as cellular debris from increasing numbers of dead cells will affect the purity of the plasmid preparation.

2. A substitute complex medium for growth of *E. coli* cultures has been used by many investigators to obtain higher yields of plasmid DNA from mini-preps. This medium was named TB for "Terrific Broth," and is a buffered, enriched culture medium that enables bacteria to grow to higher densities than is usually observed with growth in LB. Just as with LB broth, care should be taken not to allow cultures to overgrow. The recipe for 1 liter of TB is as follows: dissolve 11.8 g of peptone, 23.6 g of yeast extract, 9.4 g of $K_2HPO_4$, and 2.2 g of $KH_2PO_4$ in 950 ml of distilled $H_2O$. Add 4 ml of glycerol and autoclave to sterilize.

## Day 3

1. Harvest cultures of *E. coli* LE392 (pRY121) (obtained from instructor) by centrifugation at top speed (2500–3000 rpm) in a Beckman TJ-6 tabletop centrifuge (or equivalent) for 5 minutes at room temperature.

2. Decant the culture supernatant and discard. Add 100 $\mu$l of solution I to each tube and resuspend the cell pellet by vortexing (minimize introduction of bubbles, if possible). Transfer this suspension to a 1.5-ml microcentrifuge tube.

3. Add 200 $\mu$l of solution II to each sample. Immediately invert gently 10 times to facilitate mixing the viscous material.

4. Add 200 $\mu$l of solution III to each tube. Mix by inversion (five times) and place on ice for 5 minutes.

5. Add 500 $\mu$l of 5 *M* LiCl. Mix by inversion (five times) and spin the tubes in a microcentrifuge for 10 minutes at top speed at

room temperature to pellet protein, RNA, and chromosomal DNA contaminants. *Note:* As an optional step, 2–3 μl of chloroform may also be added to the precipitating mixture, after addition of LiCl but prior to centrifugation, in order to facilitate the removal of particulate matter (mostly protein) from the resulting supernatant after centrifugation.

6. For each sample, remove 900 μl of the supernatant and transfer it to a fresh 1.5-ml microcentrifuge tube. *Avoid carryover of any particulate matter.* Add 600 μl of isopropanol and mix by inverting the tubes approximately 10 times.

7. Centrifuge at top speed in a microcentrifuge for 10 minutes at room temperature. Carefully aspirate or decant the isopropanol supernatant and discard.

8. Wash the precipitated nucleic acid pellet by adding 1 ml of ice-cold 70% (v/v) ethanol, mixing the tubes by gently inverting three to five times, and centrifuging at top speed in a microcentrifuge for 2 minutes at room temperature. Aspirate or decant the ethanol supernatant and discard. Rest all tubes in an inverted position on an absorbent paper towel for about 5–10 minutes to allow residual ethanol to evaporate.

9. Resuspend each pellet in 50 μl of TE buffer. Add 1 μl of RNase (10 mg/ml). Place the tubes in a 55°C heat block for 15 minutes to facilitate dissolution of DNA. Mix occasionally by vortexing. RNase is an extraordinarily stable enzyme and is not affected by the conditions of high heat.

## Day 4

10. The presence of plasmid pRY121 DNA may be confirmed by restriction digestion and/or agarose gel electrophoresis of these samples, as outlined previously in Exercise 8. Approximately 2 μl of this mini-prep DNA is usually sufficient to visualize by ethidium bromide (EtBr) fluorescence following electrophoresis. A comparison of either digested or undigested samples from this

exercise to pRY121 maxi-prep DNA (from Exercise 6) would allow for verification of the size and/or restriction pattern of the vector.

## References

Birnboim, H. C. (1983). A rapid alkaline extraction procedure for the isolation of plasmid DNA. *Methods Enzymol.* **100**, 243–255. [This reference complements the previous reference on alkaline lysis cited in Exercise 4 (Birnboim and Doly, 1979) and represents a more detailed description of this technique.]

Hanahan, D. (1983). Studies on transformation of *E. coli* with plasmids. *J. Mol. Biol.* **166**, 557–580. [A reference outlining the various parameters and factors that influence transformation of *E. coli*.]

He, M., Wilde, A., and Kaderbhai, M. A. (1990). A simple single-step procedure for small-scale preparation of *Escherichia coli* plasmids. *Nucleic Acids Res.* **18**, 1660. [A report describing the use of LiCl in the preparation of plasmid DNA. This modification of the alkaline lysis procedure allows for rapid isolation of plasmid without using phenol or chloroform.]

Ward, A. C. (1990). Single-step purification of shuttle vectors from yeast for high frequency back-transformation into *E. coli. Nucleic Acids Res.* **18**, 5319. [A description of the basic technique for plasmid DNA isolation from yeast; it cites an increased efficiency in successful recovery of plasmids shuttled into *E. coli*.]

## Questions

1. Explain what would be required to calculate the transformation efficiency of *E. coli* in this exercise. List five factors that affect transformation efficiency and propose reasons for the low level of transformation observed in this exercise.

2. Describe the different regions of DNA on pRY121 that enable it to replicate and grow in both *E. coli* and *S. cerevisiae*.

3. Draw a flow chart outlining the various steps of the alkaline lysis mini-prep procedure. List the purpose of each step and reagent/ solution used toward obtaining plasmid DNA from bacterial cells.

# 13

# Protein Assays

## Introduction

With this exercise, we start a series of experiments designed for familiarization with some basic tools of protein biochemistry. The determination of the concentration of protein in a solution is a critical analytical procedure that must be accomplished rapidly, accurately, with high sensitivity, and at low cost. A variety of procedures have been developed to quantitate protein in biological specimens. Methodologies continue to be improved and new assays developed. It is thus difficult to decide which assay(s) to use for a sample. The choice of assay is often based on individual idiosyncrasies of laboratory experience or bias in a particular subfield of biology. In this exercise, we become acquainted with two assays for protein: the Lowry assay and the Bradford assay. By comparing and contrasting these assays, you may evaluate their relative merits and limitations. The real goal is for you to be prepared to continually adapt proper analytical measurements to your particular biological problem.

## A. Lowry Assay

The Lowry reaction for protein determination is an extension of the biuret procedure. The first step involves formation of a copper–protein complex in alkaline solution. This complex then reduces a phosphomolybdic–phosphotungstate reagent to yield an intense blue color. This assay is much more sensitive than the biuret method but is also more time consuming. The precaution to be observed when performing this assay concerns addition of the Folin reagent. This reagent is stable only at acidic pH; however, the reduction indicated above occurs only at pH 10. Therefore, when the Folin reagent is added to the alkaline copper–protein solution, mixing must occur immediately so that the reduction can occur before the phosphomolybdic–phosphotungstate (Folin) reagent breaks down.

Phenols are capable of reducing molybdenum in a complex of phosphomolybdotungstic acid (Folin–Ciocalteu reagent). The tyrosine residues of protein provide phenolic groups and cupric ions enhance the sensitivity. When treated with Folin–Ciocalteu reagent, proteins produce color (blue) in varying degrees depending on their tyrosine content. Hence different proteins give different color values. Generally, the method can be used to determine 20 to 200 $\mu$g of protein per milliliter.

## B. Bradford Protein Assay

The Bradford protein assay is a dye-binding assay based on the differential color change of a dye in response to various concentrations of protein. The dye reagents are commonly purchased from Bio-Rad (Richmond, CA). In some research applications, this assay is recommended as a replacement for other protein assays, especially the widely used Lowry method, for several reasons. First, the Bradford protein assay is much easier to use. It requires one reagent and 5 minutes to perform as compared to the three reagents and 30–40 minutes typical for the Lowry assay. Second, because the absorbance of the dye–protein complex is relatively stable, the Bradford assay does not require the critical timing necessary for the Lowry assay. Third, the Bradford assay is not affected by many of the compounds that limit the application of the Lowry assay.

The Bradford protein assay is based on the observation that the absorbance maximum for an acidic solution of Coomassie Brilliant Blue G-250 shifts from 465 to 595 nm when binding to protein occurs. The extinction coefficient of a dye–albumin complex solution is constant over a 10-fold concentration range. Thus Beer's law may be applied for accurate quantitation of protein by selecting an appropriate ratio of dye volume to sample concentration. Over a broader range of protein concentrations, the dye-binding method gives an accurate, but not entirely linear, response.

# PART A   LOWRY PROTEIN ASSAY

## Reagents/Supplies

Bovine serum albumin (BSA) solutions (0.2, 0.4, 0.6, 0.8, 1.0 mg/ml)
Folin–Ciocalteu reagent: dilute 1 : 1 with distilled $H_2O$ so that its final concentration is 1 $N$
Glass test tubes (12 × 75 mm)
Micropipettor
Reagent A: 1.56 g of $CuSO_4 \cdot 5H_2O$/100 ml of distilled $H_2O$
Reagent B: 2.37 g of sodium tartrate/100 ml of distilled $H_2O$
Reagent C: 2.0 g of NaOH + 10 g of $Na_2CO_3$/500 ml of distilled $H_2O$
Reagent D: Combine and mix 100 ml of reagent C + 1 ml of reagent A + 1 ml of reagent B, in this order
Solution of protein of *unknown* concentration (between 100 and 2000 $\mu$g/ml)

## Equipment

Spectrophotometer
Vortex

### *Instructor's Note* _____

An exercise to demonstrate the effect of various potential contaminants on protein determinations is to add the following to tubes containing 4 ml of a 0.5-mg/ml BSA solution: 80 $\mu$l of 10% (w/v) sodium dodecyl sulfate (SDS), 40 $\mu$l of 20% (w/v) sucrose, 80 $\mu$l of 1 $M$ Tris (pH 8), and 16 $\mu$l of 0.5 $M$ EDTA (pH 8). A comparison of assay results obtained from these solutions to those from a standard curve of BSA alone is instructive. See Appendix 19, Effects of Common Contaminants of Protein Assays.

## Procedure

1. Place 1 ml of reagent D in appropriately labeled tubes.

2. Add 100 $\mu$l of each sample of BSA to a tube containing reagent D, add various amounts of protein solution of unknown concentration to tubes of reagent D, and vortex *immediately*. Be sure to have a tube containing no BSA as blank. Incubate for 10 minutes at room temperature.

3. After the 10-minute incubation, add 100 $\mu$l of Folin–Ciocalteu reagent to the samples and vortex *immediately*. Incubate for 30 minutes at room temperature.

4. Read the samples at $OD_{660}$ on the spectrophotometer, using reagent D as the blank.

5. Plot a curve of the $OD_{660}$ (subtract the blank $OD_{660}$ reading from each $OD_{660}$ for the BSA solutions) versus milligrams of BSA per milliliter. This standard curve is then used to determine the protein concentration of the unknown.

## PART B    BRADFORD ASSAY

## Reagents/Supplies

Bovine serum albumin solutions (0.2, 0.4, 0.6, 0.8, 1.0 mg/ml)
Bradford dye (Bio-Rad)
Brown glass bottle
Glass funnel
Micropipettor
Unknown samples
Whatman paper (No. 1)

## Equipment

Spectrophotometer
Vortex

## Procedure

1. Dilute 1 volume of the Bradford dye with 4 volumes of distilled water.

2. Filter the dye through Whatman No. 1 paper and store at room temperature in a closed brown glass bottle.

3. Add 2.5 ml of dilute dye to appropriately labeled tubes.

4. Add 50 $\mu$l of the BSA and the unknown samples to tubes containing the dilute dye and vortex immediately.

5. Read the samples at $OD_{595}$.

6. Plot the resulting standard curve.

7. Use the standard curve to determine the concentration of the unknown sample.

## References

Bradford, M. M. (1976). A rapid and sensitive method for the quantitation of microgram quantities of protein utilizing the principle of protein dye binding. *Anal. Biochem.* **72**, 248–254. [This paper describes a simple, rapid, and sensitive method for protein determination.]

Coligan, J., Dunn, B., Ploegh, H., Speicher, D., and Wingfield, P. (1995). "Current Protocols in Protein Sciences," Vol. 1. John Wiley & Sons, New York. [Available in looseleaf and CD-ROM formats, this text is a comprehensive practical compilation of methods in protein biochemistry.]

Lowry, O. H., Rosebrough, N. J., Farr, A. L., and Randall, R. J. (1951). Protein measurement with Folin–phenol reagent. *J. Biol. Chem.* **193**, 265–275. [This is the classic paper on protein determination. It has been cited more often than any other paper in the biochemical literature.]

## Question

1. Compare and contrast the Lowry and Bradford protein assays with regard to compatibility with various potential contaminating compounds.

# 14

# Qualitative Assay for β-Galactosidase in Yeast Colonies

## Introduction

Although quantitative analyses yielding specific activities are often preferred, there are times when a facile qualitative determination of enzymatic activity is sufficient to attain specific experimental goals. β-Galactosidase (*lacZ*) is commonly used as a reporter gene to detect gene expression in cells, microorganisms, and in some cases sectioned animals. The ability to rapidly detect colorimetric changes via substrate cleavage by this enzyme is responsible for its universal use as a molecular tag for gene expression. The use of β-galactosidase as a reporter enzyme in *Saccharomyces cerevisiae* to study a variety of cellular processes including secretion, receptor-mediated signal transduction, and protein–protein interaction is strengthened by the lack of an endogenous β-galactosidase activity in *S. cerevisiae*.

The plasmid pRY121 contains the *Escherichia coli* gene, *lacZ*, encoding the enzyme β-galactosidase. On pRY121, *lacZ* is under the control of the yeast *GAL1–10* promoter (see plasmid map in Exercise 8). This promoter enables the induction of β-galactosidase by the addition of galactose to the growth medium. Raffinose is used as a carbon source to provide cellular growth in the place of glucose, which represses *GAL10*-mediated *lacZ* expression. The presence of the enzyme can be readily measured by a colorimetric procedure using an indicator substrate called X-Gal. This compound yields a blue color on its hydrolysis, catalyzed by β-galactosidase.

This exercise employs a filter assay using intact yeast for the rapid evaluation of β-galactosidase activity *in situ*. Yeast cells transformed with pRY121 can be grown under appropriate conditions to induce expression, then transferred to filter paper, permeabilized by a quick freeze–thaw with liquid nitrogen, immersed in a buffer containing a chromogenic substrate, and incubated until color development is detected. This procedure enables the qualitative characterization of transformants for the presence of β-galactosidase activity and is amenable to the processing of multiple samples simultaneously.

## Reagents/Supplies

5-Bromo-4-chloro-3-indolyl-β-D-galactopyranoside    (X-Gal),    2% (w/v): dissolved in N,N-dimethylformamide; prepare fresh prior to use

Ethanol (70% v/v), for sterilization of forceps

Forceps

Liquid nitrogen

Replica plating velvets, sterilized by autoclaving; wrapped in foil (ReplicaTech, Inc.)

*Saccharomyces cerevisiae* YNN281 from a fresh streak plate

*Saccharomyces cerevisiae* YNN281 (pRY121) from a fresh streak plate or recent transformation

Toothpicks, sterile

Styrofoam packaging container with lid

Whatman No. 3 paper, circles cut out to fit inside a Petri plate

Whatman No. 50 paper, circles cut to fit inside a Petri plate, sterilized by autoclaving; wrapped in foil

YEPD agar plates (see Exercise 2)

YEPGal agar plates [2% (w/v) peptone, 1% (w/v) yeast extract, 2% (w/v) galactose, 2% (w/v) raffinose, 2% (w/v) agar]

Z buffer: To 800 ml of distilled $H_2O$, add 16.1 g of $Na_2HPO_4 \cdot 7H_2O$, 5.5 g of $NaH_2PO_4 \cdot H_2O$, 0.75 g of KCl, 0.25 g of $MgSO_4 \cdot 7H_2O$, and 2.7 ml of 2-mercaptoethanol; adjust the pH to 7.0 and bring to 1 liter with distilled $H_2O$. *Do not autoclave*

## Equipment

Incubator, set at 30°C
Liquid nitrogen container
Replica plating block (ReplicaTech, Inc.)

## Procedure

### Day 1

1. Using sterile toothpicks, take a small patch (about the size of a small fingernail) from each streak plate of the *S. cerevisiae* strains YNN281 and YNN281 (pRY121) and place on *both* the YEPGal and the YEPD agar plates. Use the same source colonies for both the YEPGal *and* YEPD plates. Place the patches at least 2 cm apart from each other near the center of each plate.

2. Invert and incubate the plates at 30°C for 1–2 days (until the next laboratory period). Dense patches of cells should be present by the end of this time.

### Day 2

3. Place circular sterilized Whatman No. 50 filter paper directly onto new YEPGal and YEPD agar plates, using ethanol-sterilized forceps. Allow the media to completely soak into the filter paper (this can be facilitated by pressing down with the forceps).

4. Mark one edge of the YEPGal and YEPD plates (both sets) to indicate orientation. Carefully, but quickly, to avoid contamina-

tion, center a sterile velvet on top of a replica plating block and secure it.

5. Remove the lid and invert the YEPGal plate containing the yeast patches. Place it on top of the sterile velvet while checking the orientation of the plate (by lining up the mark on the side of the petri plate with the mark on the replica plating block). Gently tap the bottom of the plate to ensure contact with the velvet.

6. Remove the plate from the velvet and replace it with the YEPGal plate onto which the filter paper was placed. Again, check that the orientation of the plates is maintained. Tap the bottom of this plate to aid the transfer of cells to the filter paper.

7. Remove the plate containing the inoculated filter from the velvet and incubate it at 30°C for 1–2 days (until the next laboratory period). Repeat Steps 4–7 to transfer YEPD cell patches to the YEPD plate with filter paper on it.

## Day 3

8. Prepare a color development chamber by placing a piece of circular Whatman No. 3 paper into the lid of a petri plate. Soak this filter with 2 ml of Z buffer containing 20 $\mu$l of 2% (w/v) X-Gal. Place the bottom of the petri plate onto the inverted lid containing a completely soaked filter. *Perform the following steps for each individual filter with colonies, then repeat for subsequent filters.*

9. Using forceps (it is not necessary to work aseptically at this point), peel the filter paper from the original plates (the paper should now have cells grown on top of it).

### Safety Note
Avoid contact with liquid nitrogen. It is painfully cold!

10. Carefully immerse the filters in liquid nitrogen (in an appropriate container), using forceps, for about 10–15 seconds. This freezes

the cells and paper and the subsequent thawing results in permeabilization of the cells.

11. Place the frozen filter directly on top of the Z buffer-soaked filter (Step 8), cover with the bottom of the plate, and incubate at 30°C overnight to allow for color development. Cells expressing β-galactosidase should exhibit a light blue color.

# References

Breeden, L., and Nasmyth, K. (1985). Regulation of the yeast *HO* gene. *Cold Spring Harbor Symp. Quant. Biol.* **30,** 643–650. [This report is the original reference in which this β-galactosidase filter assay was employed.]

Fields, S., and Song, O. (1989). A novel genetic system to detect protein–protein interactions. *Nature (London)* **340,** 245–246. [This paper is a report of a genetic assay that uses yeast cells to detect protein–protein interactions *in vivo*. The β-galactosidase filter assay used in this exercise has now been incorporated by other investigators to improve the detection scheme employed in this paper.]

Guarente, L. (1983). Yeast promoters and *lacZ* fusions designed to study expression of cloned genes in yeast. *Methods Enzymol.* **101,** 181–191. [An excellent technical review of the use of β-galactosidase gene fusions in yeast to study gene expression.]

# Questions

1. Draw the structural formula for X-Gal and show which portion of the molecule is responsible for the blue color.

2. On returning to the laboratory after a weekend, you find all filters, including those containing cells grown on YEPD and YEPGal, are intensely blue. Explain this observation.

# 15

# Determination of
# β-Galactosidase in
# Permeabilized Yeast Cells

## Introduction

Cell permeabilization techniques are often useful for many applications relating to enzyme technology. For example, permeabilization procedures are usually rapid and do not destroy cellular enzymes. Thus, the total amount of an enzyme associated with a cell can be assayed after permeabilization. A number of permeabilization methods for yeast have been developed, such as use of detergents, organic solvents, and desiccation.

In this exercise, *Saccharomyces cerevisiae* cells will be treated with a combination of an organic solvent (toluene) and a detergent (sarkosyl or sodium lauroyl sulfate), which effectively dissolves the permeability barriers (membrane lipids), allowing free access of added substrates to intracellular proteins. Because the cell wall barrier is not broken down by this treatment, most proteins remain entrapped and associated with the cell.

## Reagents/Supplies

N-Lauroylsarcosine (5%, w/v)
Microcentrifuge tubes

Micropipettor

o-Nitrophenyl-β-D-galactopyranoside (ONPG), 4 mg/ml of distilled H₂O

Overnight cultures of *Saccharomyces cerevisiae* YNN281 (pRY121) and *Saccharomyces cerevisiae* YNN281

SF solution: 0.85% (w/v) NaCl, 3.7% (v/v) formaldehyde

Sodium carbonate (Na₂CO₃): 1 $M$

Toluene

YNB without amino acids plus supplements, as described below (Step 1), plus galactose (see Exercise 2 for preparation of YNB)

Z buffer (see Exercise 14)

## Equipment

Incubator set at 37°C
Incubator set at 28°C
Microcentrifuge
Spectrophotometer
Vortex

## Procedure

1. Grow *S. cerevisiae* YNN281 (pRY121) and *S. cerevisiae* YNN281 at 30°C in 50 ml of YNB, plus tryptophan (30 $\mu$g/ml), lysine (30 $\mu$g/ml), adenine sulfate (30 $\mu$g/ml), histidine (30 $\mu$g/ml), galactose (3%, w/v), and raffinose (2%, w/v) to mid-log (50 Klett units or $OD_{600}$ of 0.7, which is about $1.7 \times 10^7$ cells/ml). Be sure to add uracil (30 $\mu$g/ml) to the YNN281 culture, but omit this supplement from the YNN281 (pRY121) culture.

2. Harvest (microcentrifuge for 1 minute at top speed at room temperature) 3.0 ml of each culture in microcentrifuge tubes, discard the supernatant, combine like pellets, and resuspend the pellets in 1 ml of Z buffer. Wash the pellets by centrifugation again at top speed at room temperature and resuspend in 300 $\mu$l of Z buffer.

3. Remove 100 $\mu$l of resuspended cells and add to 900 $\mu$l of SF solution in cuvette. Read the $OD_{600}$ with a spectrophotometer.

4. Place 150 $\mu$l of the remaining cell suspension in a fresh microcentrifuge tube. Add 1.5 $\mu$l of toluene and 1.5 $\mu$l of N-lauroylsarcosine (5%, w/v). Vortex at top speed several times.

5. Evaporate the toluene from the solution for 30 minutes at 37°C on a shaker with caps off the tubes.

6. Add 20 $\mu$l of toluenized cells to 1.5 ml of a prewarmed (28°C) reaction mixture of 1 ml of ONPG (4 mg/ml) and 0.5 ml of Z buffer. More cells can be added if equivalent amounts of Z buffer are left out. Vortex and start timing. When the color is yellow (or after 120 minutes), add 0.5 ml of $Na_2CO_3$ (1 $M$) to stop the reaction.

7. Read the optical density at 420 and 550 nm. The $OD_{550}$ corrects for light scattering by the cell suspension. Alternatively, the reaction mixture can be spun to pellet the cells and the $OD_{420}$ of supernatant read.

8. The units of activity can be calculated:

$$\text{Units} = \frac{1000 \times (OD_{420} - 1.75\ OD_{550})}{t \times v \times OD_{600}}$$

where $t$ is the assay time (minutes), $v$ is the volume of cells used in the assay (milliliters), $OD_{600}$ is the cell density at the start of the assay, $OD_{420}$ is the combination of absorbance by o-nitrophenol and light scattering by cells, and $OD_{550}$ is the light scattering by cells.

# References

Cordeiro, C. A. A., and Freire, A. P. (1994). Protein determination in permeabilized yeast cells using the Coomassie Brilliant Blue dye binding assay. *Anal. Biochem.* **223**, 321–323. [The paper presents a method to assay proteins in permeabilized cells in order to compare enzyme activities obtained with permeabilization techniques to those with homogenates or purified enzymes.]

Gowda, L. R., Joshi, M. S., and Bhat, S. G. (1988). *In situ* assay of intracellular enzymes of yeast *(Kluyveromyces fragilis)* by digitonin permeabilization of cell membrane. *Anal. Biochem.* **175**, 531–536. [This paper presents an alternative method for yeast permeabilization, using the detergent digitonin.]

Reynolds, A., and Lundblad, V. (1989). Yeast vectors and assays for expression of cloned genes. *In* "Current Protocols in Molecular Biology" (Ausubel, F. M., *et al.,* eds.). John Wiley & Sons, New York, pp 13.6.2–13.6.4. [A description of yeast cell permeabilization by organic solvent/detergent mixture is presented.]

## Questions

1. What are the advantages and disadvantages of enzyme assays on permeabilized cells versus cell extracts?

2. Why is galactose the major carbon source in the growth medium for this experiment?

# Assay of β-Galactosidase in Cell Extracts

## Introduction

A critical step in enzyme purification is the extraction of cellular material from whole cells. The preparation of cell extracts (sometimes called cell-free extracts) must be done with care to ensure complete release of the enzyme from cellular material without denaturation of the enzyme itself. Often a compromise is made in extraction between complete cell destruction and enzyme recovery, because methods used to break open cells often are harmful to proteins.

A variety of methods are used for cell extract preparation: sonication, homogenization with a glass or Teflon pestle, grinding with aluminum or glass beads, and high pressurization and rapid release of pressure (see Table 16.1).

For yeast, the most convenient and effective method for cell extract preparation involves grinding with glass beads. The shearing action of the glass beads during vortexing causes the rupture of the yeast cell. The very thick and tough cell wall of yeast makes methods such as sonication, osmotic lysis, and homogenization ineffective in breaking the yeast cell.

## Reagents/Supplies

Bradford protein assay reagents (see Exercise 13, Part B)
Conical glass tubes (15 ml)

Glass beads (200-$\mu$m diameter, acid washed; Sigma)
Glass beads (Braun Biotech, 0.45–0.50 mm)
$o$-Nitrophenyl-$\beta$-D-galacto pyranoside (ONPG): 4 mg/ml
Phenylmethylsulfonyl fluoride (PMSF): 100 m$M$; dissolve in 100%
    isopropanol immediately before use
*Saccharomyces cerevisiae* YNN281 (pRY121), log-phase culture
Sodium carbonate (Na$_2$CO$_3$): 1 $M$
Z buffer (see Exercise 14)

**Table 16.1   Cell Extraction Methods**

| Method | Bacteria | Yeast | Mammalian cells |
|---|---|---|---|
| Sonication | Useful | Not effective | Useful |
| French press | Useful | Useful only at very high pressures (20,000 psi) | Not useful |
| Osmotic lysis | Useful for release of periplasmic enzymes | Not effective | Useful |
| Homogenization with glass/Teflon pestle | Not effective | Not effective | Very useful |
| Grinding with glass beads | Useful | Most effective | Not useful |

# Equipment

Centrifuge
Hemocytometer
Microscope
Spectrophotometer
Vortex
Water bath, 28°C

## Instructor's Note

Grow a 250-ml culture of *S. cerevisiae* YNN281 (pRY121) as described in Exercise 15 (Step 1) for harvest at class time at $1 \times 10^7$ cells/ml.

## Procedure

1. Harvest 250 ml of *S. cerevisiae* YNN281(pRY121) at late exponential growth phase ($1 \times 10^7$ cells/ml) by centrifugation at 3000 $g$ for 10 minutes at room temperature or at 4° C. [The number of cells should be determined by $OD_{600}$ or hemocytometer counting prior to harvest (see Appendix 14, Determination of Cell Number). An $OD_{600}$ reading of 0.4 = $1.0 \times 10^7$ cells/ml.]

2. Resuspend the cells, using conical tubes, to 2 ml ($5 \times 10^8$ cells/ml) in Z buffer containing 1 m$M$ PMSF. Add glass beads to the height of the meniscus of the cell suspension.

3. Vortex the cells with glass beads at high speed for 30 seconds and cool on ice for 30 seconds. Repeat the vortexing procedure and cool the cells for six cycles.

4. Let the beads settle, remove and save the cell lysate, add 2 ml of buffer (Z buffer plus 1 m$M$ PMSF) to the beads, mix, remove the lysate, and combine the lysates. Check the percentage of lysis by determining the cell number by hemocytometer. Centrifuge the lysate at 5000 $g$ for 20 minutes at 4° C. Collect the supernatant as the cell extract.

5. Determine the protein concentration (Exercise 13, Part B, Bradford Assay) of the recovered extract and perform the enzyme assay as described below (steps 6–11).

6. Dilute an aliquot of the lysate to 0.1 mg of protein/ml of Z buffer.

7. Add 100 $\mu$l of the diluted sample to 0.9 ml of Z buffer and incubate at 28°C for 5 minutes. Run a blank of 1.0 ml of Z buffer only, and a blank of diluted sample, but add Z buffer instead of ONPG at Step 8.

8. Begin the reaction by adding 0.2 ml of ONPG (4 mg/ml) that has also been equilibrated to 28°C.

9. Incubate the sample at 28°C until a faint yellow color has developed. If color develops immediately after adding the ONPG, dilute the extract or use less extract and start over from Step 7 above.

10. Stop the reaction by adding 0.5 ml of 1 $M$ $Na_2CO_3$, and record the length of time of incubation.

11. Read the $OD_{420}$.

12. Calculate the specific activity (SA) by using the following equation (see Exercise 15, Step 8):

$$SA \text{ (units/mg)} = \frac{OD_{420} \times 380}{\text{time (minutes)} \times \text{protein (mg)}}$$

## References

Guthrie, C., and Fink, G. R. (1991). "Guide to Yeast Genetics and Molecular Biology." Academic Press, San Diego, California. [This is a comprehensive compilation of procedures for studying yeast. The book is reprinted from Vol. 194 of "Methods in Enzymology."]

Pringle, J. R. (1975). Methods for avoiding proteolytic artifacts in studies of enzymes and other proteins from yeasts. In "Methods in Cell Biology XII: Yeast Cells" (Prescott, D., ed.). Academic Press, New York, pp. 149–184. [The review describes practical procedures for inhibiting proteases in yeast extracts.]

Silverman, S. J. (1987). Current methods for *Saccharomyces cerevisiae*. I. Growth. *Anal. Biochem.* **164**, 271–277. [An excellent introduction to practical techniques for the growth and manipulation of *S. cerevisiae*.]

## Questions

1. Compare the specific activity of β-galactosidase obtained from permeabilized cells (Exercise 15) and cell extracts (this exercise). Account for differences that may exist between results from these protocols.

2. What is the purpose of adding PMSF to the Z buffer (see Pringle, 1975).

# 17

# β-Galactosidase Purification

## Overview

The purification of a protein is essential for a detailed study of its structure and function. In essence, the particular properties of a protein are utilized to separate it from the numerous other proteins in the cell. A purification scheme is usually devised to separate molecular species on the basis of selective solubility, chromatographic behavior, and molecular weight. In cases where these are not sufficient to obtain purification, more sophisticated techniques must also be employed. At every stage of a purification it is important to know the "specific activity" of an enzyme—the enzymatic activity per milligram of protein, which indicates the degree of purity of the enzyme.

A large-scale culture of *Saccharomyces cerevisiae* strain YNN281 (pRY121) is harvested, homogenized by use of glass beads, and centrifuged to remove the cell debris. Ammonium sulfate is added to this supernatant to precipitate most of the β-galactosidase and results in a two- to fivefold purification. The precipitate is redissolved in buffer and the ammonium sulfate is dialyzed out against the same buffer. The sample is applied to a gel filtration column followed by elution. The activity peaks are pooled and the purity of the protein is determined by electrophoresis on sodium dodecyl sulfate (SDS)-polyacrylamide gels. The sequence used for the purpose of this course is given in Figure 17.1. Other purification steps (beyond the scope of this introductory course) that are often employed include ultrafil-

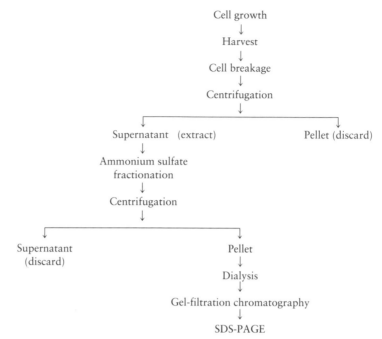

**Figure 17.1**   Flow chart for the purification of $\beta$-galactosidase.

tration and chromatography such as ion-exchange, affinity, and high-performance liquid chromatography.

## PART A   GEL-FILTRATION CHROMATOGRAPHY: COLUMN CALIBRATION

### Introduction

Prior to chromatography of the cell extract, it is necessary to prepare and calibrate the gel-filtration column. Thus we start the exercise on enzyme purification with column calibration and return to chromatography of the extract following the ammonium sulfate precipitation.

Gel-filtration chromatography, also called size-exclusion chromatography, is based on the principle that the flow of molecules is impeded due to the differences in their hydrodynamic volumes. Separation of molecules is determined by the specific gel-filtration medium used and hydrodynamic volume of the molecule. One of the common materials used for gel-filtration chromatography is Sephadex, which is a modified dextran cross-linked to give a hydrophilic, three-dimensional network. The degree of swelling is an important characteristic of the gel, in which the matrix is a minor component. Loose gels are used for fractionation of high molecular weight substances, whereas more compact gels are used for separation of lower molecular weight compounds.

The choice of an appropriate gel-filtration matrix depends on the molecular size and chemical properties of the substances to be separated. Each particular type of Sephadex provides separation mainly within a particular molecular weight range determined by the degree of swelling of the gel. Molecules above a certain molecular weight are totally excluded from particular gels and are eluted in a volume approximately equal to the void volume (volume of interstitial liquid between gel grains in the gel bed). The relationships between elution behavior and molecular properties are given in manufacturer applications catalogs.

Gel-filtration yields are usually excellent, as there is little or no retention of substances on column material. A common application of gel filtration is in the analysis of mixtures of molecules of different molecular weight. Thus, the determination of molecular weights is a common application of gel-filtration chromatography. Other preparative applications include the use of this gel material for desalting, which refers to the removal of salts and other low molecular weight compounds from solutions of macromolecules, concentration of high molecular weight substances, exchange of buffers, removal of organic material (such as phenol from preparations of nucleic acids), removal of low molecular weight, radioactively labeled compounds from solutions of labeled proteins, termination of the reaction between macromolecules and low molecular reactants, and removal of products or inhibitors from enzymes. Furthermore, gel filtration can be used to study chemical equilibria and to separate cells and particles. The student is advised to read gel-filtration applications catalogs supplied by the various manufacturers.

## Reagents/Supplies

Blue Dextran: Average molecular weight, 2,000,000 (Sigma)
Molecular weight standards (Bio-Rad or Sigma)
Sephadex G-200 (Pharmacia)
Syringe
Z buffer (see Exercise 14): Degas for 5–10 minutes prior to use

## Equipment

Column (2.5 × 50 cm, 250-ml volume) with stand and 500-ml reservoir; constant pressure flask
Fraction collector
Peristaltic pump
Ultraviolet (UV) monitor or spectrophotometer

### Instructor's Note _____

Suspend 6 g of Sephadex G-200 in 500 ml of Z buffer in a 1-liter beaker. Avoid excess stirring that may break beads. Do not use a magnetic stirrer. Place in a boiling water bath for 5 hours or incubate at 20°C for 72 hours. Stir occasionally with a glass rod.

## Procedure

### Day 1

1. Decant ~60% of the supernatant from above the hydrated Sephadex G-200.

2. Mount the column vertically on a suitable stand. Attach a 500-ml reservoir to the top of the column.

3. Close the column outlet. Add 50 ml of Z buffer. To remove air bubbles, inject Z buffer by syringe into the outlet tubing until it passes through the column bed.

4. Carefully pour the Sephadex G-200 suspension into the reservoir. The reservoir enables all the suspension to be poured in a single operation.

5. Allow the gel to settle.

6. Open the column outlet and, using a peristaltic pump, adjust the flow rate to not greater than 1 ml/min.

7. Run at least two bed volumes of degassed Z buffer through the gel-filtration bed in order to allow the bed to stabilize.

8. Poured columns should be protected from contaminating microbial growth during storage. Sodium azide ($NaN_3$) added to the buffer to 0.02% (w/v) is often used in cases of long-term storage.

**Day 2**

9. Application and elution of samples is accomplished as follows.

   a. Close the outlet and remove most of the eluent above the gel surface.

   b. Open the outlet to allow drainage of eluent. *Do not allow the bed to run dry.*

   c. To measure the void volume, layer Blue Dextran gently on top of the bed.

   d. Open the column outlet to allow the sample to enter into the gel.

   e. Wash the sample remaining on the bed surface and column wall into the bed with a small amount of Z buffer.

   f. Refill the column with eluent and connect to the pump. In the absence of a pump, the flow rate can be controlled by hydrostatic pressure, by use of a constant pressure flask (see manufacturer applications manuals).

   g. Collect fractions in a suitable volume (i.e., 4-ml fractions every 10 minutes) for assays. This volume must be determined empirically and will differ according to the experimental purification involved. To determine the void volume, measure the volume collected until Blue Dextran elutes.

h. Repeat the elution (Steps 9c–g, above) with molecular weight markers. If colored markers are used, simply determine the fraction number or elution volume (in milliliters) where the marker elutes from the column. If noncolored markers are used, read the column eluent by $OD_{280}$ with a UV monitor or read each fraction at $OD_{280}$ with a spectrophotometer. Plotting the molecular weight of the marker proteins versus elution volume yields a calibration curve for the column.

## PART B    AMMONIUM SULFATE SALTING-OUT OF β-GALACTOSIDASE AND COLUMN CHROMATOGRAPHY

## Introduction

In cell extracts, the desired protein is often less than 1% of the total protein. In this case, differential solubility procedures are generally used as the first step to separate a protein of interest from "contaminating" proteins. Differential solubility procedures have a high capacity to eliminate gross impurities. Methods of higher resolution are used in subsequent purification steps.

Heat treatment, selective precipitation (precipitation of protein at its p$I$ in a solution of pH = p$I$), organic solvent precipitation, polyethylene glycol precipitation, and salt fractionation are all differential solubility techniques. However, the most useful in the majority of cases is salt fractionation with ammonium sulfate. Ammonium sulfate is highly soluble in water, available at high purity and low cost, and stabilizes proteins.

At low ionic strengths, addition of ammonium sulfate has a "salting-in" effect of enhancing protein solubility. With addition of more ammonium sulfate, protein becomes less soluble resulting in a "salting-out" or precipitation of the protein. The concentration of ammonium sulfate at which each protein precipitates is characteristic for that protein and varies with pH, temperature, and protein concentration. Proteins precipitated by ammonium sulfate usually retain native conformation, can be redissolved readily, and can be stored in the precipitated form.

### Instructor's Note

For this exercise it will be necessary to prepare a large-scale cell extract of *S. cerevisiae* YNN281 (pRY121). Follow the procedures in Exercise 15, using 2–4 liters of cell culture that has been induced to produce β-galactosidase by growth in galactose. It is advisable to review the basic rules for handling enzymes (Appendix 18) prior to this exercise.

## Reagents/Supplies

Ammonium sulfate [$(NH_4)_2SO_4$, as pure as possible]: Molecular biology grade
Beakers
Cell extract from YNN281 (pRY121) culture (prepared by instructor)
Dialysis tubing (see Instructor's Note, below)
Materials for β-galactosidase assay (see Exercise 16)
Materials for protein assay (see Exercise 13)
Phenylmethylsulfonyl fluoride (PMSF) (see Exercise 16)
Z buffer (see Exercise 14)

## Equipment

Calibrated columns (see Part A of this exercise)
Centrifuge
Cold room or ice bath
Magnetic stirrers

### Instructor's Note

Prepare dialysis tubing as in Appendix 1, Exercise 6A. For this exercise more porous tubing, such as one with an MW 25,000 cutoff, can be used; thus dialysis times can be relatively short. For small samples (as little as 50 $\mu$l can be dialyzed), very rapid equilibrium is attained owing to the large surface area available.

## Procedure

### Day 1

1. Dilute the cell extract to 100 ml with Z buffer in a 250-ml beaker. Place it on a stir plate in an ice bath. Add a stirrer to the extract solution and rapidly add PMSF to a 1 m$M$ final concentration. Remove a small amount of extract for enzyme and protein assay and for sodium dodecyl sulfate-polyacrylamide gel electrophoresis (SDS-PAGE) (see Part C of this exercise).

2. Slowly add enough solid ammonium sulfate (13.4 g) to make a solution 25% (w/v) in ammonium sulfate. You may wish to consult a table of fractionation with ammonium sulfate (Segal, 1976). Continue stirring for 10 minutes, then centrifuge at 10,000 $g$ for 10 minutes at 4°C.

3. Decant and save the supernatant; note the volume. Dissolve the precipitate in Z buffer (about 1–2 volumes of precipitate). It is desirable to keep the protein as concentrated as possible to improve stability and to prevent excessive dilution. Material that does not dissolve is probably particulate or denatured protein and can be removed and discarded after centrifugation (1000 $g$, 5 minutes at 4°C).

4. Assay dilutions of the redissolved precipitate and the supernatant for $\beta$-galactosidase activity and for protein content.

5. Continue in this mode to take precipitates in saturation ranges of 25–40% (add an additional 8.4 g of ammonium sulfate), 40–60% (add 12.0 g of ammonium sulfate), 60–80% (add 12.9 g of ammonium sulfate), and 80% supernatant. As $(NH_4)_2SO_4$ may lead to acidification of the solution, add NaOH if required to keep the solution in a neutral range (pH 6.5–7.5).

6. Prepare a table, using Table 17.1 as an example, and fill in the data points (indicated by a ?)

**Table 17.1    Purification of β-Galactosidase**

| Saturation range (%) | β-Galactosidase precipitated (%) | Protein precipitated (%) | Purification factor[a] |
|---|---|---|---|
| 0–25 | ? | ? | ? |
| 25–40 | ? | ? | ? |
| 40–60 | ? | ? | ? |
| 60–80 | ? | ? | ? |
| 80 | ? | ? | ? |

[a] Specific activity of fraction/specific activity of original cell activity.

7. Dissolve the ammonium sulfate precipitate, containing the majority of enzymatic activity, in Z buffer and dialyze against 1 liter of Z buffer overnight at 4°C, with one buffer change to remove excess salt. Save 100 $\mu$l of dialyzed extract for SDS-PAGE (see Part C of this exercise).

## Days 2 and 3

8. Apply the dialyzed sample to the gel-filtration column as described in Part A of this exercise. (It is preferable to perform this chromatography in a cold room or in a refrigerated chamber.) Elute and collect fractions, using the same procedures as in column calibration. Assay fractions for protein and β-galactosidase activity. Pool and save the fractions with maximum activity and store at 4°C.

## PART C    SODIUM DODECYL SULFATE-POLYACRYLAMIDE GEL ELECTROPHORESIS

## Introduction

Analysis of proteins by gel electrophoresis with respect to purity and molecular weight is a common procedure in most biotechnology laboratories. Modern adaptations of the technique of sodium dodecyl sulfate-polyacrylamide gel electrophoresis (SDS-PAGE) facilitate the study of such proteins in a rapid and efficient manner.

The porosity of a gel formed as a result of the copolymerization of acrylamide with the cross-linker bisacrylamide acts as a molecular sieve in which macromolecules of various size and charge can be separated. In the presence of the negatively charged detergent, sodium dodecyl sulfate (SDS), and a reducing agent, commonly 2-mercaptoethanol, disulfide bonds in and between polypeptides are broken and the resulting protein subunits are bound in the form of micelles by SDS, thereby providing all molecules bound with an overall negative charge. This serves as a method by which the varying charges of the specific amino acid residues of different polypeptides are eliminated as a factor in the electrophoretic separation of these molecules, resulting in a means of differentiation based, in general, on the molecular weight of the molecules. Therefore, on application of an electric field through an SDS-containing protein separation gel, the migration of smaller, low molecular weight polypeptides will occur faster than that of larger, high molecular weight polypeptides. Depending on the size of the molecules to be separated and relative quantities of proteins to be analyzed, the percent concentration of acrylamide can be changed, as desired, to achieve optimal results.

In this exercise, we will utilize this technique to examine the protein content of cell extracts of a genetically engineered *Saccharomyces cerevisiae* that has been induced to produce the enzyme β-galactosidase with a tetrameric molecular weight of ~460,000. The analysis of these extracts in comparison to known protein standards will provide a method by which the identification and confirmation of this protein product can be achieved rapidly.

## Reagents/Supplies

### *Safety Note* _____
Wear gloves. Acrylamide is a neurotoxin.

Acrylamide-bis: 30% (w/v) acrylamide, 0.8% (w/v) bisacrylamide; dissolve 30 g of acrylamide and 0.8 g of bisacrylamide in 80 ml of distilled, deionized $H_2O$. Once dissolved, bring to 100 ml with distilled, deionized $H_2O$. Filter through Whatman No. 1 paper. Store at 4°C in a brown glass bottle

Ammonium persulfate (100 mg/ml): Prepare fresh before use

Binder clips

Butanol, water-saturated

Destaining solution (50 ml of methanol, 860 ml of distilled, deionized $H_2O$, 90 ml of glacial acetic acid; add in this order. Mix well

Extracts from *Saccharomyces cerevisiae* YNN281 (pRY121)

    1. Crude cell extract

    2. Ammonium sulfate dialysate

    3. Gel-filtration fractions

*β*-Galactosidase, standard (from *Escherichia coli*; Sigma)

Gel combs and spacers

Glass baking dishes (8 × 12 inches, 2 quart)

Glass plates

Gloves

Hamilton syringes, 10–20 $\mu$l, if available

Micropipettor

Molecular weight standards for SDS-PAGE (Bio-Rad)

Paper towels

Pasteur pipet

Razor blades

Running buffer: 7.2 g of glycine, 1.51 g of Tris-HC1, 0.5 g of SDS for 500 ml in distilled, deionized $H_2O$

Sample buffer, 2×: 0.25 *M* Tris-HCl (pH 6.8) 20 ml of glycerol, 4 g of SDS, 2 mg of bromphenol blue, 10 ml of 2-mercaptoethanol; bring to 100 ml with distilled, deionized $H_2O$

Separating gel buffer: 1.5 *M* Tris-HCl (pH 8.8)

Sodium dodecyl sulfate (SDS): 5% (w/v)

Stacking gel buffer: 0.5 *M* Tris-HCl (pH 6.8)

Staining solution: 1 g of Coomassie Blue, 500 ml of methanol, 100 ml of glacial acetic acid, 400 ml of distilled, deionized $H_2O$; add dye to methanol, stir for 1 hour, add distilled, deionized $H_2O$

and acid, stir for 30 minutes, and filter through Whatman No. 1
filter paper
Storage containers
N,N,N',N'-Tetramethylethylenediamine (TEMED)
Vacuum flask, 50 ml

## Equipment

Constant-voltage power supply
Gel casting stand
Mini-vertical slab gel apparatus
Platform shaker
Polyacrylamide gel apparatuses (available from many manufacturers)

### Instructor's Note _____

If you plan to carry out Exercise 18 (Western Blot), be sure to run
a duplicate gel, which should not be stained.

## Procedure

### A. Assembling Gel Apparatus

Day 1

1. Clean and dry all materials before assembly and gel casting, using
   soapy water and rinsing well to remove any film, etc.

2. Assemble the gel apparatus in the gel casting stand.

### B. Pouring Gels

Day 1 (continued)

1. Prepare 10% (w/v) acrylamide separating gels by combining the
   following reagents from the prepared stock solutions in a vacuum

flask: 10 ml of acrylamide-bis solution (see Safety Note, above), 11.8 ml of distilled, deionized $H_2O$, 7.5 ml of separating gel buffer, and 0.3 ml of 10% (w/v) SDS.

2. Remove any excess air in the solution by degassing it for 5 minutes.

3. Add 300 $\mu$l of freshly prepared ammonium persulfate and 30 $\mu$l of TEMED stock, and swirl the flask to facilitate mixing without introducing any bubbles. *Caution:* This initiates the polymerization process, so be prepared to work quickly after this step.

4. Pour the gel solution into the casting apparatus (or use a Pasteur pipet) to a level ~2 cm from the top of the glass plate while being careful to avoid bubble formation.

5. Quickly, using a micropipettor, layer each gel, dropwise, with water-saturated butanol. This aids in the formation of a bubble-free interface on which to pour the stacking gel. Allow the gel to polymerize for about 20 minutes.

6. Once the gel is polymerized, pour off the butanol by carefully inverting the casting unit into a paper towel.

7. Prepare the stacking gel by combining the following solutions in a 50-ml vacuum flask: 3 ml of acrylamide-bis, 5 ml of stacking gel buffer, 11.6 ml of distilled, deionized $H_2O$, and 0.2 ml of 10% (w/v) SDS. Degas for 5 minutes.

8. Add 200 $\mu$l of ammonium persulfate and 20 $\mu$l of TEMED. Swirl and pour into the casting unit to the level of the top of the glass plates.

9. Working quickly, carefully insert the combs into the gels. While working them in, try to remove any bubbles that may exist by gently wiggling the combs in the stacker. Allow the gel to stand for ~1 hour or until it is polymerized.

10. Gently remove the gels from the casting unit. (*Note:* Gels may be stored for 2–3 days in a tightly sealed container with a little water.) Rinse the gels with water to remove excess acrylamide.

## C. Gel Electrophoresis

Day 2

1. Mount a gel in the gel apparatus.

2. Pour a little running buffer into the top half of the apparatus to check for leaks. If sealed, fill both the top and bottom of the apparatus until the gel and electrodes are covered with buffer.

3. Prepare dilutions of the three extract samples (crude, ammonium sulfate, and gel filtration) to achieve samples of 10, 20, and 40 $\mu$g of protein to load onto the gel (based on previously obtained protein concentration in the sample). Add a volume of 2× sample buffer equal to the volume of each sample. Also prepare a 20-$\mu$g sample of $\beta$-galactosidase standard and a sample of molecular weight standards for loading. (*Note:* If prestained electrophoresis standards are available, it is not necessary to dilute them with 2× sample buffer, because the progress of these standards may be visualized while running the gel.)

4. Gently remove the comb from the stacking gel.

5. Fill the wells with the appropriate samples in the desired order (take note of the loading order), using a 10 to 20-$\mu$l Hamilton syringe or equivalent. Rinse the syringe with a little running buffer between samples to avoid cross-contamination. Clean the syringe *immediately* after use with distilled water and ethanol.

6. Connect the electrodes, turn on the power supply, and set the voltage at 160 volts. Apply constant voltage until the dye front is near the end of the gel.

## D. Staining Gels

Day 2 (continued)

1. Turn off the power supply, disconnect the electrodes, and discard the buffer. Carefully remove the gel from the apparatus.

2. Remove the spacers from the sides of the gel and pry the two plates apart, using a razor blade. The stacking gel may be trimmed off at this point, if desired. [*Note:* If performing a Western blot (Exercise 18), do not stain the gel.] Place the gel in a staining solution (~100 ml) in a glass baking dish (2 quart) and agitate on a platform shaker for 15 minutes.

3. Pour off the staining solution and destain the gel in destaining solution for 10 minutes. Change the destaining solution and repeat two or three times over 2 hours, or until the desired resolution is achieved.

## References

Coligan, J., Dunn, B., Ploegh, H., Speicher, D., and Wingfield, P. (1995). "Current Protocols in Protein Sciences," Vol. 1. John Wiley & Sons, New York. [Available in looseleaf and CD-ROM formats, this text is a comprehensive practical compilation of methods in protein biochemistry.]

Green, A. A., and Hughes, W. L. (1955). Protein fractionation on the basis of solubility in aqueous solutions of salts and organic solvents. In "Methods in Enzymology," Vol. I. Academic Press, New York, pp. 67–87. [A classic reference for protein purification.]

Hames, B. D., and Rickwood, D. (1987). "Gel Electrophoresis of Proteins: A Practical Approach." IRL Press, Oxford. [This book contains much practical information on SDS-PAGE of proteins.]

Pharmacia Corporation. "Pharmacia Product Guide. Gel Filtration: Theory and Practice." Pharmacia Corporation, Uppsala, Sweden. [This manufacturer manual provides practical advice on the proper use of Sephadex in gel-filtration chromatography.]

Segal, I. H. (1976). "Biochemical Calculations," 2nd Ed. John Wiley & Sons, New York. [This book contains useful exercises and supplementary materials (e.g., ammonium sulfate fractionation tables).]

# Questions

1. Fill in the following table for your purification of β-galactosidase.

| Step | Activity (units) | Protein (mg) | Specific activity (U/mg) | Yield (%) | Purification factor |
|---|---|---|---|---|---|
| 1. Crude extract | ? | ? | ? | 100 | 1.00 |
| 2. $(NH_4)_2SO_4$ dialysate | ? | ? | ? | ? | ? |
| 3. Gel filtration pooled fractions | ? | ? | ? | ? | ? |

2. How would you verify that the band corresponding to a molecular weight of 116,000 is the monomer of β-galactosidase?

# 18

# Western Blot: Probe of Protein Blot with Antibody to $\beta$-Galactosidase

## Introduction

Proteins can be transferred by electrophoresis from various media, such as polyacrylamide gels, to membranes of nitrocellulose or nylon. The proteins are immobilized on the membrane, providing a blot that can be probed by various means. Such a blot is called a Western blot when antibodies are used to detect the transferred proteins.

Before the blot can be probed, unoccupied sites on the membrane must be saturated with protein to prevent the antibody from reacting nonspecifically with the membrane. The blot is then probed for a specific protein using primary antibodies directed against the protein of interest. Subsequently, secondary antibodies (or antibody-binding proteins, such as protein A) that recognize the primary antibody and have been conjugated to an enzyme are added. Addition of a specific substrate for the coupled enzyme yields a colored, precipitated product at the site of the protein of interest on the membrane.

Western blots are very useful in a number of applications of molecular biology. Expression of proteins can be monitored during development of an organism or in response to regulatory signals. Expression of cloned genes in heterologous cells can be detected by

Western blotting. Posttransitional processing of proteins and protein degradation can also be measured using this technique.

In this exercise we use the material generated in Exercise 17, Part C, to demonstrate this procedure. Specifically, a sodium dodecyl sulfate (SDS)-polyacrylamide gel containing $\beta$-galactosidase is probed with antibodies to this protein.

## PART A    TRANSFER OF PROTEIN TO MEMBRANE

### Reagents/Supplies

Blotting paper: Whatman 3 MM
Glass baking dishes, $15 \times 15$ cm (or large enough to accommodate the gel)
Gloves
Nylon membrane: Immobilon P (Millipore Corp.), or equivalent
Pipet
Scissors
SDS-PAGE gel (unstained) from Exercise 17, Part C
Transfer buffer: 25 m$M$ Tris, 192 m$M$ glycine [20% (v/v) methanol], pH 8.3. Dissolve 3.03 g of Tris base and 14.4 g of glycine in distilled $H_2O$, add 200 ml methanol, and adjust the volume to 1 liter with distilled $H_2O$; do not add acid or base to adjust pH

### Equipment

Semidry transfer cell (Bio-Rad Trans-Blot SD or equivalent)

### Procedure

Wear gloves throughout this exercise to avoid contaminating the blots with proteins from your hands.

1. Following electrophoresis (Exercise 17, Part C), equilibrate the gel in transfer buffer for 15 minutes at room temperature.

2. While the gel is equilibrating, cut a piece of nylon membrane to the same dimensions as the gel. Wet the membrane by slowly sliding it at a 45° angle into the transfer buffer and soaking it for 15 minutes.

3. Cut two pieces of blotting paper to the dimensions of the gel. Completely saturate the paper by soaking it in transfer buffer.

4. Assemble the blotting paper, nylon membrane, and gel in the transfer apparatus as follows:

   a. Place one piece of the presoaked blotting paper onto the transfer apparatus. Roll a glass pipet over the surface of the paper to exclude air bubbles.

   b. Place the presoaked nylon membrane on top of the blotting paper. Remove bubbles as in Step 4a.

   c. Carefully place the equilibrated gel on top of the nylon membrane. Gently remove bubbles as in Step 4a.

   d. Place the second sheet of presoaked blotting paper on top of the gel. Remove bubbles as in Step 4a.

   e. Assemble the transfer cell and plug it into the power supply according to manufacturer instructions.

5. Transfer the protein from the gel to the nylon membrane in the assembled unit for 30 minutes at 20 volts.

6. Following transfer, disconnect the unit from the power supply. Discard the blotting paper and remove the nylon membrane for further processing.

## PART B    PROBE OF PROTEIN BLOT WITH ANTIBODY TO β-GALACTOSIDASE

### Reagents/Supplies

Antibody solution: Antibody to β-galactosidase (5 Prime → 3 Prime, Inc.) diluted 1:100 into TTBS/milk (1 g of nonfat, dry milk in 10 ml of TTBS)

Blot of proteins on nylon membranes from Part A of this exercise

3,3'-Diaminobenzidine solution: 0.5 mg/ml in 0.05 $M$ Tris (pH 7.6): Add $H_2O_2$ [30% (v/v) solution], 10 $\mu$l/10 ml of diaminobenzidine, just before use. (*Caution:* Diaminobenzidine is a carcinogen. Dispose of properly as hazardous waste.) Alternatively, 4-chloro-1-naphthol can be used

Filter paper

Glass baking dish

Nonfat, dry milk powder

Protein A–peroxidase solution (purchase from various sources): Dilute 1 : 500 into TTBS/milk (1 g of nonfat, dry milk in 10 ml of TTBS)

TBS (Tris-buffered saline): 20 m$M$ Tris, 500 m$M$ NaCl (pH 7.5). Add 4.84 g of Tris base to 58.48 g of NaCl and bring to 2 liters with deionized water. Adjust the pH to 7.5 with HCl

TTBS (Tween 20 wash solution): 20 m$M$ Tris, 500 m$M$ NaCl, 0.05% (w/v) Tween 20 (pH 7.5). Add 0.5 ml of Tween 20 to 1 liter of TBS

## Equipment

Agitator platform

Incubator at 37°C

## Procedure

Wear gloves throughout this exercise

1. Prepare 10% milk in TBS by adding 1 g of nonfat milk to 10 ml of TBS. Using a glass dish slightly larger than the membrane to conserve solutions, place the nylon membrane in enough milk–TBS to cover the membrane. Agitate slightly on a platform agitator for 30 minutes at 37°C.

2. Decant and discard the milk. Add the TTBS and incubate for 5 minutes at room temperature. Repeat the wash two more times.

3. Add a minimum volume (enough to allow movement of membrane during agitation) of antibody solution. Incubate with agitation for 1 hour at 37°C. Wash three times (5 minutes for each wash) with TTBS.

4. Add a minimum volume of protein A–peroxidase solution. Incubate with agitation for 1 hour at 37°C. Wash three times, 4 minutes each, with TTBS.

5. Add $H_2O_2$ to 3,3'-diaminobenzidine solution (see Reagents/Supplies). Add a minimum volume of this diaminobenzidine/$H_2O_2$ solution (enough to wet the membrane). Incubate at room temperature and watch for color development, which should be completed within 5–10 minutes.

6. Rinse with distilled water to stop the reaction. Place the membrane on filter paper to air dry.

## References

Blake, M. S., Johnston, K. H., Russell-Jones, G. J., and Gotschlich, E. C. (1984). A rapid sensitive method for detection of alkaline phosphatase-conjugated anti-antibody on Western blots. *Anal. Biochem.* **136**, 175–179. [This reference describes the use of antibodies and enzyme–antibody complexes for detection of immobilized proteins.]

Gershoni, J., and Palade, G. M. (1983). Protein blotting: Principles and applications. *Anal. Biochem.* **131**, 1–15. [This article contains information on the theory and practical uses of the Western blot.]

Towbin, J., Staehelin, T., and Gordon, J. (1979). Electrophoretic transfer of proteins from polyacrylamide gels to nitrocellulose sheets: Procedure and some applications. *Proc. Natl. Acad. Sci. U.S.A.* **76**, 4350–4354. [This paper describes the first example of the electrophoretic transfer of proteins from gels to membranes.]

## Questions

1. What is the purpose of equilibrating the gel in the transfer buffer before performing the blot?

2. Suggest a means by which efficient transfer of proteins from the gel can be monitored.

3. What is the purpose of the milk in the transfer buffer?

4. Describe the reactions taking place during detection of the protein on the membrane.

# Alternative Protocols and Experiments

*Exercise 1A*

Isolation and Characterization of Auxotrophic Yeast Mutants

*Exercise 2A*

Measurement of pH

*Exercise 3A*

Use of the Spectrophotometer

*Exercise 6A*

Isolation of Plasmid DNA: The Maxi-Prep

*Exercise 10A*

Colony Hybridization

# Isolation and Characterization of Auxotrophic Yeast Mutants

## Introduction

*Saccharomyces cerevisiae* is widely recognized as an ideal eukaryotic microorganism for biotechnology applications such as production of proteins, hormones, and other pharmaceutically important compounds. Yeast have many technical advantages as microorganisms such as stable haploid and diploid phases of the life cycle, rapid growth, successful expression of heterologous genes, and the ease of mutant isolation. Furthermore, *S. cerevisiae* has been successfully employed in all phases of genetics, such as mutagenesis, recombination, and regulation of gene action. The ability to generate mutants of desired phenotypes is an important tool for biotechnological applications such as increasing the amount of biosynthesis of a particular compound or producing an altered enzyme with more desirable characteristics.

Because spontaneous mutation frequencies are low, yeast is usually treated with such mutagens as ultraviolet radiation, nitrous acid, ethylmethane sulfonate (EMS), diethyl sulfate, and 1-methyl-1-nitro-nitrosoguanidine (MNNG) in order to enhance the frequency of mutants. These mutagens are remarkably efficient and can induce mutations at a rate of $5 \times 10^{-4}$ to $10^{-2}$ per gene without substantial killing. Even though there are known methods to increase the proportion of mutants by killing off the nonmutant with nystatin and other agents, it is usually unnecessary to use selective means to obtain reasonable yields of mutants. In this experiment, auxotrophic mutants will be isolated from MNNG-treated yeasts.

Auxotrophic mutants have been invaluable for the elucidation of biochemical pathways as well as for the study of the relationship between enzyme structure and function. Studies on the intermediates accumulated by amino acid auxotrophs have facilitated the unraveling of biochemical pathways.

The standard method described here involves treatment of wild-type yeast strains with MNNG. After mutagenesis, the strain is diluted and plated on a complete medium plate at a concentration giving about 100–200 cells per petri plate. After these cells have grown into colonies, they are transferred to various media by replica plating. Auxotrophic mutants are detected by lack of growth on a minimal medium which contains glucose, potassium phosphate, ammonium sulfate, a few vitamins, salts, and trace metals. The specific requirements of auxotrophic mutants can be determined by testing the colony from the original YEPD plate on various types of synthetic media. Putative mutants are placed on a YEPD plate according to a pattern. After they are fully grown, they are then transferred to minimal medium plates containing pools of various amino acids, purines and pyrimidines, and other metabolites, each at a final concentration of about 20 $\mu$g/ml. From the pattern of growth of a particular strain, it is possible to identify the specific requirement of the mutant strain (see Table A-1.1A).

Generally, a colony will respond on a plate containing one of the pools from 1–5 and on another plate containing one of the pools from 6–9, thus allowing for direct identification of a single growth factor requirement. For example, a colony growing on pools 1 and 7 requires histidine and a colony growing on 3 and 8 requires trypto-

**Table A-1.1A    Pools for Medium Plates**

| Pools | Pools | | | | |
|---|---|---|---|---|---|
| | 1 | 2 | 3 | 4 | 5 |
| 6 | Adenine | Guanine | Cysteine | Methionine | Uracil |
| 7 | Histidine | Leucine | Isoleucine | Valine | Lysine |
| 8 | Phenylalanine | Tyrosine | Tryptophan | Threonine | Proline |
| 9 | Glutamate | Serine | Alanine | Aspartate | Arginine |

phan. If a colony grows on only one of the nine pools, it requires more than one of the nutrients in that pool.

## Reagents/Supplies

Amino acid pools (see Introduction to this exercise, and Instructor's Note, below)

Dilution blanks, sterile (9.9 and 9.0 ml)

N-Methyl-N'-nitro-N-nitrosoguanidine (MNNG): 0.1 M in sodium phosphate buffer (pH 7). *Caution:* This is a powerful mutagen

Replica plating block (RepliTech, Inc.)

*Saccharomyces cerevisiae* X2180-1A (see Appendix 6), or any wild-type strain

Test tubes with 8 ml of sterile sodium thiosulfate (5%, w/v)

Toothpicks, sterile

Velveteen pads, sterile

YEPD plates (see Exercise 2)

YNB plates containing amino acid pools (see Exercise 2)

## Equipment

Incubator at 30°C

### *Instructor's Notes*

1. One day before the start of the experiment inoculate the entire surface of a YEPD slant with S. *cerevisiae* X2180-1A. Incubate at 30°C overnight.

2. On the day of class, there should be ~7.5 × 10$^8$ cells on slants. Suspend the cells in 1.5 ml of sterile sodium phosphate (pH 7.0) and add 0.7 ml of cell suspension to 1 ml of buffer. Add 50 μl of nitrosoguanidine (MNNG) and agitate.

3. Prepare the amino acid pools (see Appendix 10 and the introduction to this exercise). Add to the YNB medium so that the final

concentration in YNB plates is 20 $\mu$g/ml for each amino acid, purine, or pyrimidine.

## Procedure

### Day 1

1. Shake the cell suspension containing MNNG at 30°C for 1 hour. Remove 100 $\mu$l of cells. Add to 100 $\mu$l of sodium thiosulfate to inactivate the MNNG. Determine the cell density with a hemocytometer while the cells are incubating (see Appendix 14). The MNNG treatment will cause approximately 40% killing.

2. Transfer 0.2 ml of the treated cell suspension to 8 ml of sterile 5% (w/v) sodium thiosulfate.

3. To obtain 100–200 viable cells per plate, dilute the treated cells, which are in the sodium thiosulfate solution, by a factor of $10^{-4}$ with sterile water (two sequential 1:100 dilutions can be made by twice pipetting 0.1 ml of the previous dilution into 9.9 ml of sterile water or another appropriate dilution, depending on the number of cells in the original suspension, and accounting for the dilutions made during the experiment).

4. Spread either 0.1, 0.2, or 0.4 ml each on separate YEPD plates, using 10 plates for each of the 3 different volumes plated. Incubate all of the plates at room temperature (23°C) for 3 to 4 days.

### Day 3 or Day 4

5. Choose 10 or more YEPD plates containing 25 to 200 colonies per plate for the isolation of auxotrophic mutants. Transfer the colonies from each of the YEPD plates by replica plating to one YNB plate and then to one YEPD plate (see Exercise 14). Be certain that each plate is numbered and has an orientation symbol on the back. For detection of auxotrophic mutants, incubate the YNB plates and the YEPD replicas at 30°C for 1 day.

## Day 5

6. Compare each of the 10 minimal plates to the corresponding YEPD plates. Transfer colonies whose replicates have failed to grow on the minimal plates with toothpicks to a YEPD plate in a pattern. This plate serves as the master plate for Step 7. There should be approximately 5% of such colonies. Incubate the plates at 30°C overnight.

## Day 6

7. Replica plate the master plate containing the auxotrophic mutants to the nine petri plates containing the various pools of metabolites. Incubate at 30°C overnight.

## Day 7

8. Record the growth response and assign a number for each mutant.

## References

Maga, J. A., and McEntee, K. (1985). Response of *S. cerevisiae* to *N*-methyl-*N'*-nitro-*N*-nitrosoguanidine: Mutagenesis, survival and *DDR* gene expression. *Mol. Gen. Genet.* **200**, 313–321. [This paper describes the MNNG mutagenesis procedure in detail.]

Snow, R. (1966). An enrichment method for auxotrophic yeast mutants using the antibiotic "nystatin." *Nature (London)* **211**, 206–207. [This method provides a means to decrease the nonmutagenized cells from a mutagenized population.]

_____

# Measurement of pH

## Introduction

Maintenance of pH by the use of inorganic or organic buffering systems is a necessary part of any protocol designed for the study of biological systems. This laboratory exercise will demonstrate the variation in pH that can be obtained with an inorganic buffer system and how pH is measured both electronically with a pH meter and visually with a pH-sensitive dye.

pH is measured in a number of different ways. An accurate and practical method for measuring pH involves the use of a pH meter. The pH meter is a potentiometer that measures the potential developed between a glass electrode and a reference electrode. In modern instruments, the two electrodes are frequently combined into one electrode, known as a combination electrode.

The glass electrode contains a glass bulb constructed of very thin, special glass that is permeable to hydrogen ions. As a result, a potential develops across this glass membrane. The potential is linearly related to the pH. Standardization against a buffer of known $H^+$ concentration is required because the concentration of $H^+$ inside the bulb of the glass electrode changes with time (see Appendixes 2 and 3, Buffer Solutions and Preparation of Buffers and Solutions). Adjustments for temperature are necessary because the relationship between measured potential and pH is temperature dependent. In actuality, there are a number of other potentials that develop in this system (e.g., liquid junction potential, asymmetry potential) but these are usually constant and relatively independent of pH, so that the linearity of measured potential with pH is maintained.

### Instructor's Note _____
This exercise may be dispensable, depending on the level and experi-

ence of the students. However, we have found this exercise useful for all students undergoing orientation and familiarization with the laboratory environment.

## Reagents/Supplies

Beakers (250 ml)
Bromthymol blue (0.25%, w/v)
pH indicator paper
Pipets (1, 5, and 10 ml)
Potassium phosphate, dibasic (dipotassium phosphate, $K_2HPO_4$)
Potassium phosphate, monobasic (monopotassium phosphate, $KH_2PO_4$)
Standard buffer, pH 4.01
Standard buffer, pH 7.00
Test tubes, 18 × 150 mm
Volumetric flasks, 100 ml
Wash bottle

## Equipment

pH meter

### *Instructor's Note* _____
For the novice student, reading the appropriate background material in Appendixes 2 and 3 is essential for understanding buffer preparation.

## Procedure

### *Part A.  Visual Estimation of pH*

1. Prepare 0.1 *M* solutions (100 ml) of $K_2HPO_4$ and $KH_2PO_4$.

2. Set up a series of twelve 18 × 150 mm test tubes as shown in Table A-1.2A.

**Table A-1.2A    Phosphate Buffers**

| Tube number | 0.1 $M$ K$_2$HPO$_4$ (ml) | 0.1 $M$ KH$_2$PO$_4$ (ml) | pH expected |
|:-----------:|:-------------------------:|:-------------------------:|:-----------:|
| 1 | 0.3 | 9.7 | 5.30 |
| 2 | 0.5 | 9.5 | 5.59 |
| 3 | 1.0 | 9.0 | 5.91 |
| 4 | 2.0 | 8.0 | 6.24 |
| 5 | 3.0 | 7.0 | 6.47 |
| 6 | 4.0 | 6.0 | 6.64 |
| 7 | 5.0 | 5.0 | 6.81 |
| 8 | 6.0 | 4.0 | 6.98 |
| 9 | 7.0 | 3.0 | 7.17 |
| 10 | 8.0 | 2.0 | 7.38 |
| 11 | 9.0 | 1.0 | 7.73 |
| 12 | 9.5 | 0.5 | 8.04 |

3. To tubes 1, 3, 5, 7, 9, and 11, add 5 drops of bromthymol blue; mix. You now have a series of color standards covering the pH range of 5.30 to 7.73. Record your observations. This exercise simply illustrates that indicator dyes may be useful as pH indicators.

## Part B. Measurement of pH with a pH Meter

1. Standardize the pH meter, using the standard pH 7.0 buffer. Rinse the electrodes, using a wash bottle. Do not wipe the electrodes with tissue because this creates a static electric charge on the electrodes and may cause erroneous readings. Remove the last drop of water by carefully touching a piece of clean tissue paper to the drop.

2. Measure the pH of the standard pH 4.01 buffer. Reset the pH meter, if necessary. It is most important to measure the pH with two standard buffers to ensure that the pH meter is functioning properly over the entire pH range.

3. Measure the pH of the six solutions in tubes 2, 4, 6, 8, 10, and 12 (prepared in Part A) with the pH meter. Rinse the electrodes

between readings and handle them carefully. You may also wish to use pH indicator paper to get an idea of the pH of the solutions. This rapid method is often accurate enough for some applications and is especially useful for very small volumes or radioactive solutions.

4. Record your observations from Parts A and B. Correlate the measured values from Part B to the expected pH value from Table A-1.2A.

## References

Gueffrey, D. E. (1986). "Buffers: A Guide for the Preparation and Use of Buffers in Biological Systems." Calbiochem Biochemicals, San Diego, California.

Segal, I. H. (1976). "Biochemical Calculations," 2nd Ed. John Wiley & Sons, New York. [This book contains many exercises enabling mastery of concepts and procedures of pH measurements and buffer preparation.]

## Questions

1. Show your calculations for preparing the following solutions: 200 ml of 20% (w/v) NaOH, 1 liter of 1.0 $M$ Tris (MW 121.1 g/mol), and 100 ml of 0.2 $M$ EDTA (MW 372.2 g/mol).

2. How much of the above Tris and EDTA solutions is used to prepare 100 ml of TE buffer (10 m$M$ Tris and 1 m$M$ EDTA)?

3. Describe the relationship between buffer working range and its p$K$ value.

4. Discuss the term *buffer capacity*.

# Use of the Spectrophotometer

## Introduction

The spectrophotometer is utilized by molecular biologists for accurate preparation and analysis of many types of samples. This exercise is designed to familiarize students with this instrument.

Spectrophotometers have varied applications in the qualitative analysis of sample purity, DNA and protein quantitation, cell density measurements, and assays involving enzyme-catalyzed reactions. Spectrophotometry is based on the simple premise that various compounds will differentially absorb specific wavelengths of light in either the ultraviolet (UV, 200–400 nm), visible (VIS, 400–700 nm), or near-infrared (near-IR, 700–900 nm) range. Photometric assays may directly measure sample absorbance at a given wavelength or indirectly measure an enzymatic reaction product or related binding substance that absorbs light in amounts directly proportional to the absorbance of the target compound (such as assays for protein concentration). Nevertheless, all spectrophotometers employ the same basic structural components designed to detect these variations between absorption wavelengths and densities.

The general components common to most spectrophotometer systems include a light source, wavelength selector, fixed or adjustable slit, cuvette, photocell, and analog or digital readout (Figure A-1.3A). The light source (specific for either UV or visible ranges) will emit light that is passed through the wavelength selector, usually a prism, diffraction grating, or set of screening filters, where a specific wavelength of "monochromatic" light is selectively generated (defined by its maximum emission at this wavelength). This light is then directed toward a thin slit (usually adjustable) to regulate its relative intensity before it passes through a cuvette containing the sample of interest. Cuvettes differ with respect to their absorbance characteristics and

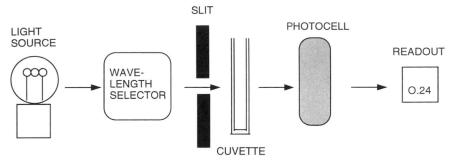

**Figure A-1.3A.** Diagram of components of spectrophotometers.

one must be careful and consistent when making measurements. Finally, absorbance is detected by a photocell in which electrons in the refracted light generate an electrical current that can be amplified and measured to yield an absorbance value, or optical density (OD), for the given sample. The OD reading is corrected against a blank, which contains all of the reagents in the experimental sample except the test compound, to obtain a true measure of the optical density of the sample.

The data obtained from spectrophotometric analyses are important for determining experimental parameters and analyzing results, especially in conjunction with data from other analytical techniques.

## Reagents/Supplies

Bromphenol blue (1.25%, w/v)
Cuvettes (alternatively, colorimeter tubes if needed)
Micropipettors and tips
Pipets, 10 ml
Test tubes, 18 × 150 mm

## Equipment

Spectrophotometer (if not available, colorimeter can be used)
Vortex

**Instructor's Note** ──────────────────────────────────

It will be important to demonstrate to the students the proper use of instrumentation used in your laboratory for this class. This exercise can be done using a variety of instruments to measure optical density. You may wish to have the students compare values obtained from different instruments.

## Procedure

1. Warm up (about 20 minutes) the spectrophotometer set at 540 nm or a Klett colorimeter (green filter, wavelength range of 500 to 570 nm) before use. The Klett reading can be converted to optical density (OD) by multiplying the Klett reading by 0.002.

2. Place 10 ml of distilled water in each of eight test tubes.

3. Use micropipettors to add to each successive tube the following amounts of bromphenol blue (1.25%, w/v): 0.5, 1, 2, 4, 10, 20, 50, and 100 $\mu$l. Manufacturer instructions for use of micropipettors should be followed scrupulously (see Appendix 5, Use of Micropipettors).

4. Vortex each tube until the dye is in solution.

5. Set the spectrophotometer/colorimeter to zero with distilled water.

6. Transfer the above dye solutions from least concentrated to most concentrated into the same cuvette or Klett tube from reading to reading.

7. Record the readings and graph the results.

## References

Brown, S. B. (1980). "An Introduction to Spectroscopy for Biochemists." Academic Press, New York.

Harris, D. A., and Bashford, C. L. (1987). "Spectrophotometry and Spectrofluorimetry: A Practical Approach." IRL Press, Washington, D.C.

## Questions

1. What may account for the difference in OD values obtained using a Klett colorimeter with a green filter as compared to a spectrophotometer set at 540 nm?

2. Consistency of micropipettor usage depends on strict attention to what operational procedures?

3. Explain the relationship among absorbance value, optical density, and percent transmittance.

# Isolation of Plasmid DNA: The Maxi-Prep

## Introduction

In practice, clones of transformed bacteria putatively containing a plasmid of interest can be identified by restriction digests of DNA from mini-preps. After identification, large-scale isolation and purification of plasmid DNA is the next step for recovering milligram amounts of plasmid. The traditional method for large-scale plasmid isolation and purification involves CsCl gradient centrifugation and is presented here.

The "maxi-prep" entails three basic steps: (1) growing bacteria and amplifying the plasmid, (2) harvesting and lysing the bacteria, and (3) purifying the plasmid away from the bacterial host. Selective inhibition of host chromosomal DNA replication with chloramphenicol during the log phase will result in "amplification" of pRY121 (see Exercise 8, Figure 8.1, for a map of the plasmid). Like other plasmids used for cloning, pRY121 has a "relaxed" origin of replication; therefore the plasmid can continue replicating even after growth of the host is arrested. Hundreds of copies of plasmid per cell can be obtained by this amplification method.

In the second step, the *Escherichia coli* cells are harvested by centrifugation. The cells are then exposed to lysozyme (in the presence of sucrose for osmotic stabilization) to weaken their cell wall structure. The cells are finally lysed gently by mixing with a mild detergent–alkali solution. Lysis releases the compact, supercoiled molecules of plasmid into solution, whereas most of the larger chromosomal DNA will remain associated with cellular debris. Because plasmids are covalently closed circles, the double-stranded molecules will not be denatured by mild treatment with alkali.

In the final step, differential centrifugation will yield a pellet containing cellular debris and the majority of cellular DNA. The clear supernatant is highly enriched for plasmid. Pure plasmid is then isolated from the supernatant by CsCl density gradient ultracentrifugation. The intercalating dye, ethidium bromide, is added in saturating amounts to the gradient. The dye will bind to nucleic acids and decrease their buoyant density. The purification step again takes advantage of the circular versus linear nature of the plasmid and host DNAs. The linear *E. coli* DNA will bind more dye per unit length than intact, supercoiled plasmid DNA. Therefore, the plasmid DNA will band below other DNA types on ultracentrifugation to equilibrium. The plasmid band is harvested, ethidium bromide is removed by extraction, and the preparation is dialyzed to remove CsCl. The quality of the plasmid DNA can be tested by agarose gel electrophoresis as described in Exercise 8.

## Reagents/Supplies

Ampicillin, stock solution: 50 mg/ml
Beaker, 500 ml
Benchkote
Centrifuge tubes, disposable, 15 ml
Centrifuge tubes, polycarbonate: Oak Ridge type, 50- and 500-ml capacity (Nalgene)
Centrifuge tubes, Quick-Seal polyallomer: 13.5-ml capacity (Beckman Instruments)
Centrifuge tube rack
Cesium chloride (CsCl), molecular biology grade
Chloramphenicol, 34 mg/ml in 100% ethanol: Protect from light by storing it in a tube wrapped in aluminum foil
Cuvette
Dialysis tubing cut into 10-cm lengths (see Instructor's Note, below)
EDTA (pH 8.0): 0.5 *M*
Erlenmeyer flask, 1 liter
*Escherichia coli* LE392 transformed with pRY121 (see Appendix 7)

Ethanol, 70% (v/v)
Ethidium bromide (EtBr) stock: 10 mg/ml (see Exercise 5)
Gloves
Glycerol
Goggles (or face shields), UV blocking
Ice
Isoamyl alcohol
Isopropanol (2-propanol)
LB broth (see Exercise 1)
Lysozyme: 0.02% (w/v) in 8% (w/v) Tris–sucrose (prepare fresh)
Mineral oil (optional)
Needles (18 and 21 gauge)
Neutralizing solution (see Exercise 5)
Paper towels
Pasteur pipets
Quick-Seal tube sealer (Beckman Instruments)
Scissors
Syringes, 5 ml
TE buffer (see Exercise 4)
TE buffer, 0.1×: 1 : 10 dilution of TE buffer in sterile water
Transfer pipets, plastic, narrow tip
Tris–sucrose solution: 8.5 g of sucrose in 100 ml of 25 m$M$ Tris base
    (pH 8.0), 2 m$M$ $MgCl_2$
Triton X-100 solution: 0.4% (w/v) Triton X-100, 50 m$M$ Tris
    (pH 8.0), 63 m$M$ EDTA
Tubing clamps (optional)
Vials, cryogenic (optional)

## Instructor's Note _____

Dialysis tubing is supplied in rolls of various widths and must be cut
to convenient lengths of 15–20 cm. Tubing should be handled with
gloves. The preferred width is 20–25 cm, and the molecular weight
cutoff used is 6000–8000. To prepare dialysis tubing, fill a 1-liter
beaker with a solution of 2% (w/v) sodium bicarbonate and 1 m$M$
EDTA and bring to a boil. Add ~20 lengths of tubing and boil for
10 min. The tubing will float and should be kept submerged. Stirring

to promote submerging is discouraged because of the risk of punctures. After boiling, rinse the tubing several times with distilled water by repeated decantation. Boil for 10 minutes in 1 m$M$ EDTA. Store the tubing submerged in 0.2 m$M$ EDTA at 4°C or it will dehydrate.

## Equipment

Analytical balance
Beckman J2-21 centrifuge or equivalent (JA-10 rotor)
Beckman ultracentrifuge or equivalent (Beckman JA-20 rotor and Ti80 rotor and caps)
Hand-held UV light
pH meter
Ring stand and clamps
Shaking incubator, set at 37°C
Spectrophotometer
Top-loading balance
Water bath at 37°C

### Instructor's Note _____

One day prior to the start of the experiment, inoculate *E. coli* LE392 transformed with pRY121 from an agar plate into 50 ml of LB broth. If desired, TB broth (described in Exercise 12, Instructor's Note) may be substituted for LB to attain higher yields of plasmid. Add 50 $\mu$l of a 50-mg/ml stock of ampicillin to maintain selective pressure for antibiotic resistance. Incubate overnight (~17 hours) at 37°C with vigorous shaking for good aeration of the culture.

## Procedure

Day 1

1. Pipet 1.5 ml of culture medium into a cuvette to use as a reference for the spectrophotometer. You may wish to freeze 1.0 ml of the

culture for future reference by adding 0.5 ml of sterile glycerol to 1.0 ml of cultures in a cryogenic vial and storing it at $-70°C$ (see Appendix 8).

2. Use 2.5 ml of overnight culture as inoculum for a 1-liter flask containing 250 ml of LB broth and 250 $\mu$l of ampicillin.

3. Incubate in a water bath at 37°C with vigorous shaking. Monitor the growth of culture over the next several hours by measuring the optical density (OD) at 600 nm, using a spectrophotometer. Make an initial measurement of optical density directly after inoculation.

4. When the $OD_{600}$ = 0.4 (log phase), add 1.2 ml of chloramphenicol solution. This will arrest bacterial protein synthesis and allow the plasmids to amplify.

5. Incubate at 37°C with vigorous shaking for 15–18 hours. (Beyond this time the bacteria will die.)

### Instructor's Notes _____

1. Precool a Beckman J2-21 centrifuge or the equivalent and rotor at 4°C.
2. Use 500 ml-capacity polycarbonate centrifuge bottles or the equivalent to harvest the bacteria. Sterility of culture beyond this point is no longer critical.
3. Centrifuge the cell suspension at $\sim$4000 g in a Beckman J2-21 centrifuge (or equivalent).
4. Decant the supernatant into a waste flask to be autoclaved. Invert the tubes on a paper towel to drain. Be careful the pellet does not slide out. The pellets should be beige in appearance. If the pellets are black, the bacteria are dead. Store the pellets on ice. A small amount of dead bacteria is to be expected.

## Day 2

1. Resuspend the cell pellet (prepared by the instructor from an overnight growth in LB with chloramphenicol) in 3.5 ml of the

Tris–sucrose solution. Transfer the suspension into a 50-ml Oak Ridge polycarbonate centrifuge tube.

2. Prepare fresh 0.02% (w/v) lysozyme solution in Tris–sucrose. Store on ice. Add 0.8 ml of lysozyme solution per 250 ml of bacterial culture.

3. Add 0.8 ml of 0.5 m$M$ EDTA, pH 8.0.

4. Add 4.0 ml of the Triton X-100 solution. Invert the tube to mix.

5. Incubate on ice for 30 minutes, inverting the tube several times every 10 minutes. This allows time for lysis of bacteria. The lysate should be highly viscous and slimy.

6. Balance the tubes against each other by adding Tris–sucrose solution, if necessary. Spin at 11,500 rpm in a Beckman JA-20 rotor (or equivalent) for 30 minutes at 4°C.

7. Pour the supernatant containing the plasmid DNA into a clean 50-ml centrifuge tube. The pellet contains bacterial debris and should be discarded. Add 6.5 ml of isopropanol and mix by gently inverting 10 times.

8. Spin at 11,500 rpm in a JA-20 rotor (or equivalent) for 30 minutes at 4°C. Decant and discard the supernatant and allow the nucleic acid pellet to air dry in an inverted position on absorbent paper for 10–15 minutes. Resuspend the pellet by adding 6 ml of 1× TE buffer and gently swirling the tube repeatedly.

9. Add 1 g of CsCl per milliliter of supernatant. Add 0.1 ml of ethidium bromide (10-mg/ml stock) per milliliter of supernatant.

### Safety Notes

a. CsCl can *burn* your skin. *WEAR GLOVES!*
b. Ethidium bromide is a *carcinogen*. Handle with *care*. Ethidium bromide is broken down by treatment described in Exercise 5. A hand-held UV light can be used to scan the work area and yourself for contamination. Ethidium bromide will fluoresce orange under UV light.

c. Remember that UV light is also hazardous and exposure should be minimized. Wear goggles or a face shield.

10. Completely dissolve the CsCl by inverting the tube repeatedly. The ensuing reaction is highly endothermic. To dissolve CsCl more rapidly, the tube may be placed in a 37°C water bath and inverted every 5 minutes. Total time to dissolve is ~15–20 minutes.

11. Transfer the solution into a 13.5-ml Quick-Seal polyallomer centrifuge tube in the following manner: Remove the plunger from a 5-ml syringe, attach an 18-gauge needle, and put it into the Quick-Seal tube. Use a Pasteur pipet to transfer solution into the syringe barrel. If any air space remains in the tube, it should be filled by the addition of a solution of CsCl (1 g/ml in 1× TE buffer) until only a small bubble is left at the neck of the tube. Mineral oil may be used to fill air gaps as well.

**Safety Note** _____

Dispose of EtBr properly. See Exercise 5.

12. Remove the syringe when the tube is filled and balance this tube against a second tube by adding CsCl solution. A blank of CsCl (1 g/ml) can be made for use as a balance if there is an uneven number of tubes.

**Safety Note** _____

It is absolutely essential that centrifuge tubes be balanced to within 0.01 g of each other prior to ultracentrifugation. Use of an improperly balanced ultracentrifuge is extremely dangerous.

13. Use a heat sealer, such as a Beckman Tube Topper, to seal the tube. Your instructor will explain its use. A *small* bubble should be left in the tube after sealing.

14. Load the tubes into a Beckman Ti80 rotor, or equivalent; grease the caps and cap rotor slots, and spin for 36–40 hours at 40,000 rpm at 20°C.

## Day 4

1. Set up a ring stand with two clamps: one to hold the tube and one to hold the UV lamp.

### Safety Note _____

Position a 500-ml beaker containing EtBr neutralizing solution (Exercise 5) under the clamps. Have paper towels on hand. (Cover the work area with absorbant paper with plastic backing such as Benchkote). WEAR GLOVES AND LABORATORY COATS!

2. Gently remove the tubes from the rotor. Do *not* disturb the gradient. Place the tubes in a centrifuge tube rack.

3. Locate the upper chromosomal and lower plasmid bands under UV light. Each will fluoresce more brightly than the surrounding solution. Protein is dark purple and floats at the top of the tube. RNA is dark red and is pelleted.

4. Clamp the tube onto the ring stand. Be sure the tubes are secured in the clamp and that the ring stand is stable! Locate the bands with UV light.

### Safety Note _____

Ethidium bromide may splatter as pressure is relieved. Clean needles in neutralizing solution and dispose of them in the proper receptacle.

5. Use a 21-gauge needle to gently and carefully puncture two or three holes in the top of the tube to break the seal and allow air to enter the tube as the plasmid band is drawn off.

6. Insert a 21-gauge needle on a 5-ml syringe (bevel side up) into the tube just underneath the plasmid band (the *lower* band).

7. *Slowly* draw off the plasmid band. If the bands are observed to be drawing close together during extraction, it is still better to leave behind a small amount of plasmid than to contaminate the harvest with bacterial DNA.

8. Slowly remove the needle containing plasmid DNA. Hold a gloved finger over the hole in the tube until you can submerge it in a neutralization solution-filled beaker.

9. Transfer the plasmid to a 15-ml disposable centrifuge tube. Record the volume. Clean up the work area with neutralization solution.

### Safety Note _____

Perform Steps 10, 11, and 12 *in a fume hood* and avoid inhalation of or skin contact with isoamyl alcohol.

10. Mix by stirring with a glass rod 30 ml of isoamyl alcohol and 20 ml of sterile distilled $H_2O$ in a 50-ml polycarbonate centrifuge tube. Allow the $H_2O$ to settle to the bottom. Use the upper, saturated organic layer to extract EtBr from the DNA.

11. Add a volume of isoamyl alcohol equal to the DNA volume to the tube. Invert the tube several times and let the aqueous phase containing DNA settle to the bottom.

12. Remove the upper, bright pink organic layer with a Pasteur pipet and discard it in a waste bottle. Repeat the procedure until the aqueous phase is colorless (seven or eight times), and then repeat once more.

13. Prepare 250 ml of 0.1× TE per preparation for first dialysis.

14. Wearing gloves, rinse a piece of dialysis tubing (prepared by the instructor) in distilled water; double-knot one end. Allow ~2.5 cm of tubing/ml of DNA solution. The tubing should be cut with an ethanol-flamed pair of scissors. Clamp the tubing in front of the knot. (*Note:* Dialysis tubing clamps are available from Spectrum, Inc. The use of these clamps can alleviate problems associated with tying knots in this tubing.)

15. Transfer DNA solution into the tubing with a plastic transfer (narrow-tip, preferably) pipet. Do not insert the pipet very far

into the tubing, owing to the risk of puncture. Knot and clamp the open end as in Step 14.

16. Put the tubing with the DNA solution into 250 ml of 0.1× TE. Stir with a magnetic spin bar on a stir plate at 4°C.

**Instructor's Note** _____
Change the TE buffer periodically over the next few days. Allow a minimum of 48 hours of dialysis.

### Day 7

1. Remove the clamp from one end and cut the dialysis tubing just below the knot with an ethanol-sterilized pair of scissors. Hold onto the tubing! Invert a 15-ml disposable polypropylene tube over the opened end. Carefully invert the tubing into the tube and dispense the contents. Alternatively, a sterile transfer pipet can be used to *carefully* remove the contents.

2. Make a 1:10 dilution of the DNA using TE buffer (0.1×) in a microcentrifuge tube.

3. Read the optical density of the DNA at 260 and 280 nm against a TE blank in quartz cuvettes.

4. Calculate the DNA concentration, purity, and total yield in milligrams (see Exercise 5).

5. Store the DNA at 4°C in a microcentrifuge tube.

## References

Clewell, D. B. (1972). Nature of Col E1 plasmid replication in *Escherichia coli* in the presence of chloramphenicol. *J. Bacteriol.* **110**, 667–676. [This paper refers to the use of chloramphenicol to promote amplification of plasmid DNA in *E. coli*.]

Heilig, J. S., Lech, K., and Brent, R. (1994). Large-scale preparation of plasmid DNA. In "Current Protocols in Molecular Biology"

(Ausubel, F. M., *et al.* eds.). John Wiley & Sons, New York, pp. 1.7.1–1.7.15. [This reference contains several detailed protocols for large-scale plasmid preparation. A comparison and discussion of the various methods available is presented as well.] ·

Radloff, R., Bauer, W., and Vinograd, J. (1967). A dye-buoyant-density method for the detection and isolation of closed circular DNA in HeLa cells. *Proc. Natl. Acad. Sci. U.S.A.* **57,** 1514–1521. [This paper is the original reference for plasmid purification by CsCl gradient centrifugation.]

## Questions

1. Compare and contrast the procedures for mini-prep and maxi-prep DNA isolation.

2. Why do chromosomal and plasmid DNA separate on centrifugation in CsCl? Why does RNA separate from DNA as well?

3. Draw a flow chart depicting the stages involved in a CsCl maxi-prep. Describe the purpose of each reagent used in the course of the procedure.

# Colony Hybridization

## Introduction

Colony hybridization allows microbial clones or virus plaques to be screened by nucleic acid hybridization to indicate those clones containing a sequence of interest (Grunstein and Hogness, 1975). Colonies are transferred to filters and lysed *in situ* to denature and immobilize DNA. The transferred DNA can then be probed by a specific cloned sequence, synthetic oligonucleotide, or RNA to indicate whether the host cell contained nucleic acid that hybridized to the probe.

In this laboratory exercise, clones of *Escherichia coli* and *Saccharomyces cerevisiae* will be screened to determine the presence of β-galactosidase sequences. The probe will be the same as the biotinylated one used in Exercise 10.

## Reagents/Supplies

Agar plates (LB + amplicillin) (see Exercise 12)
Agar plates (YEPD) (see Exercise 2)
Agar plates (YNB with supplements) (see Exercise 2). *Note:* Omit uracil
Biotinylated probe and indicator system (see Exercise 10)
*Escherichia coli* LE392
*E. coli* LE392(pRY121)
Forceps
Nitrocellulose paper or nylon membrane
Paper towels
Parafilm
Plastic wrap

*Saccharomyces cerevisiae* YNN281
*S. cerevisiae* YNN281(pRY121) (see Exercise 11)
Sodium hydroxide (NaOH): 0.5 *M*
Tris-HCl (pH 7.4): 1 *M*
Tris-HCl (pH 7.4): 1 *M*, containing 1.5 *M* NaCl
Whatman 3 MM paper

## Equipment

Incubator at 37°C
Incubator at 30°C
Vacuum oven at 80°C

### Instructor's Notes

1. Grow fresh streak plates of *S. cerevisiae* YNN281 and YNN281 (pRY121) on YEPD and YNB (with supplements) at 30°C. Grow fresh streak plates of *E. coli* LE392 and LE392(pRY121) on LB and LB + ampicillin at 37°C.

2. Use sterile toothpicks to transfer colonies onto plates containing YEPD or YNB with supplements (for yeast cultures) or LB ± ampicillin (for *E. coli*) medium. Make 2- to 3-mm streaks (small patches) in a grid pattern of about 25 colonies to serve as master plates.

3. Incubate inverted plates at 37°C (*E. coli*) or 30°C (*S. cerevisiae*) for 1–2 days.

## Procedure

### Day 1

1. Place nitrocellulose or nylon filters (cut to fit petri dishes) onto agar plates. Handle the filters with forceps. Let them stand for 1–2 minutes.

2. Mark the filters and master plates in three locations with a pencil. The master plate should be sealed with Parafilm and stored at 4°C.

3. Remove the filters from the plates by peeling them off carefully; float it *colony side up* in 0.5 *M* NaOH. (About 1 ml of 0.5 *M* NaOH on plastic wrap is convenient for this step.) Be sure the filter is wet evenly and leave at room temperature for 3 minutes.

4. Blot the side of the filters exposed to NaOH onto a paper towel, and repeat Step 3 with fresh 0.5 *M* NaOH.

5. Transfer the filters to 1 ml of 1 *M* Tris-HCl, pH 7.4. Leave for 5 minutes to allow neutralization, blot, and repeat.

6. Transfer to 1 ml of 0.5 *M* Tris-HCl (pH 7.4) containing 1.5 *M* NaCl. Leave for 5 minutes, blot, transfer to dry Whatman 3 MM paper, and let dry at room temperature for 45 minutes.

7. Place the filters between two sheets of Whatman 3 MM paper and bake the filter at 80°C in a vacuum oven for 90 minutes.

8. Remove the filters and store at 4°C until the next laboratory period.

## Day 2

9. Hybridize the probe as in Exercise 10, Part D.

## References

Grunstein, M., and Hogness, D. S. (1975). Colony hybridization: A method for the isolating of cloned DNAs that contain a specific gene. *Proc. Natl. Acad. Sci. U.S.A.* **72,** 3961–3966. [The original paper describing colony hybridization.]

Hanahan, D., and Meselson, M. (1983). Plasmid screening at high density. *Methods Enzymol.* **100,** 333–342. [This paper contains many useful comments on the methodology of colony screening.]

## Question

1. Design an experiment using colony hybridization to find a colony among a transformed population that contains a plasmid with the *lacZ* gene.

# 2

# Buffer Solutions

A buffer is a solution containing a mixture of a weak acid (HA) and its conjugate base ($A^-$) that is capable of resisting substantial changes in pH on the addition of small amounts of acidic or basic substances. Addition of acid to a buffer leads to a conversion of some of the $A^-$ to the HA form; addition of base leads to a conversion of some of the HA to the $A^-$ form. As a result, addition of either acid or base leads to a change in the $A^-$/HA ratio. In the Henderson–Hasselbalch equation, which applies to buffer systems, the $A^-$/HA ratio appears as a logarithmic function. Hence, changes in the $A^-$/HA ratio lead to only minor changes in pH within the working range of the buffer.

The working range of a buffer is determined by the p$K$ value of the system. Specifically, it falls within $\pm 1$ pH unit from the p$K$ value. Beyond that range, not enough is present of both of the buffer forms to allow the buffer to function effectively when either acid or base is added. Buffer capacity depends on the volume of the buffer and on the concentrations of the two buffer components. Buffer capacity is usually defined as the number of equivalents of either $H^+$ or $OH^-$ required to change the pH of a given volume of buffer by one pH unit.

A buffer may be prepared in two ways: (1) known amounts of the $A^-$ and HA forms may be mixed and diluted to volume; or (2) to a known amount of the HA form ($A^-$ form), a known amount of base (acid) may be added and the mixture diluted to volume. Note that the molarity of a buffer always refers to the total concentration of the buffer species. Thus, a 0.5 $M$ $H_2PO_4^-/HPO_4^{2-}$ buffer is one in which the sum of the concentrations of $H_2PO_4^-$ and $HPO_4^{2-}$ is 0.5 mol/liter.

## Henderson–Hasselbalch Equation

The Henderson–Hasselbalch equation is derived from the dissociation (ionization) reaction for a weak acid (HA):

$$HA = H^+ + A^-$$

where HA is a Brönsted acid (proton donor, conjugate acid) and A is a Brönsted base (proton acceptor, conjugate base). The two species HA and A are said to form a conjugate acid–base pair. The equilibrium (dissociation, ionization) constant for this reaction ($K$ or $K_{eq}$) is given by

$$K = \frac{[II^+]\,[A^-]}{[HA]}$$

where brackets indicate molar concentration (concentration expressed in terms of moles per liter). By taking logarithms and rearranging, the Henderson–Hasselbalch equation is obtained:

$$pH = pK + \log \frac{[A^-]}{[HA]}$$

where pH equals $-\log[H^+]$ and $pK$ equals $-\log K$.

As indicated, the pH is defined as

$$pH = -\log[H^+] = \log \frac{1}{[H^+]}$$

This definition (due to Arrhenius) was chosen in order to obtain small positive numbers for the hydrogen ion concentration. The pH scale usually runs from 0 to 14, with 7 representing neutrality, pH values above 7 representing basic conditions, and pH values below 7 representing acidic conditions. A change of 1 unit in the pH represents a 10-fold change in the hydrogen ion concentration.

In pure water, and in dilute solutions, the product of the hydrogen ion and hydroxide ion concentrations (known as the ion product of water, $K_w$) is a constant:

$$[H^+][OH^{-1}] = K_w = 10^{-14} \quad (\text{at } 25°C).$$

# 3

# Preparation of Buffers and Solutions

When making up solutions, observe the following guidelines:

1. Use the highest grade of reagents, when possible.

2. Prepare all solutions with the highest quality distilled water available.

3. Autoclave solutions when possible. If the solution cannot be auto-claved and you wish to store it, sterilize by filtration through a 0.22-$\mu$m filter.

4. Check the pH meter carefully, using freshly prepared solutions of standard pH, before adjusting the pH of buffers.

5. Always label the container with the name of the solution, the percentage or concentration, your name/course, and date. For highly basic solutions, such as 1 $M$ NaOH, be sure to use plastic containers as glass is corroded by bases.

6. Store solutions cold when possible.

Time can be saved by making up concentrated stock solutions (e.g., 1 $M$ Tris) that can be used to make up a range of different solutions.

## Percentage Solution

A percentage solution is one in which the exact concentration of the solute is known in 100 ml of a liquid or solution. The concentration may be expressed or determined by weight or volume; it may be written out in grams and milliliters and expressed as follows:

Percentage (w/v) = weight (in grams) in 100-ml volume of solution
Percentage (v/v) = volume (in milliliters) in 100-ml total volume of solution

### Examples by Weight

*For a 1% (w/v) aqueous solution of sodium chloride,* weigh out 1 g of sodium chloride (NaCl) and add to 100 ml of distilled water, or 10 mg of NaCl per milliliter.

*For a 5% (w/v) aqueous solution of sodium chloride,* weigh out 5 g of sodium chloride and add to 100 ml of distilled water.

*For 1000 ml of a 5% (w/v) aqueous solution of sodium chloride,* use 10 times the above amounts, that is, 50 g of sodium chloride in 1000 ml of distilled water.

### Example by Volume

The percentage of the solution is stated on the label, and a percentage solution should be determined from this value. For example, the percentage on a bottle of commercial strength hydrochloric acid varies from 36 to 40%. If you want a 1% solution of HCl, it is erroneous to take 1 ml of 36–40% HCl and add it to 99 ml of distilled water.

An acceptable and correct formula for determining the amount of 36% (v/v) HCl needed to make up 100 ml of a 1% (v/v) aqueous solution of HCl is as follows:

$$36 \times X = 1 \times 100$$
$$36X = 100$$

$$X = \frac{100}{36}$$

$X = 2.78$ ml of 36% (v/v) HCl in 97.22 ml of distilled
water

See Appendix 4 for common commercial strengths of acids and bases.

## Dilutions for Solutions (Percentage by Volume)

The most accurate formula for making dilutions of solutions is the following:

Percentage (%) you have × unknown volume (ml)
= percentage (%) you want × volume (ml) wanted

### *Example of Preferred Method*

To make up 1000 ml of 70% (v/v) ethanol from 95% (v/v) ethanol, substitute the known amounts in the above formula:

$95 \times X = 70 \times 1000$

$95X = 70,000$

$$X = \frac{70,000}{95}$$

$X = 736.8$ ml of 95% (v/v) ethanol in 263.2 ml of distilled water

Although the above method is preferred, you may use a generally accepted method of subtracting the percentage required from the percentage strength of the solution that is to be diluted. The difference will be the amount of water that is to be used.

### *Example of Subtraction Method*

The percentage required is 70% (v/v) ethanol and the solution to be diluted is 95% (v/v) ethanol. As $95 - 70 = 25$, 70 parts of 95% (v/v) ethanol and 25 parts of distilled water are the amounts to be combined. To make up approximately 1 liter of 70% (v/v) ethanol, you would simply place 700 ml of 95% ethanol in a 1000-ml graduated cylinder and fill to the 950-ml mark with distilled water.

## Molar Solutions

A molar (mole, molecular) solution is one in which 1 liter (1000 ml) of the solution contains the number of grams of the solute equal to its molecular weight (sum of atomic weights).

### *Examples of Molar Solutions*

*To make up a 1 M solution of NaCl (sodium chloride):*

1. Obtain the atomic weights of the elements (see a Periodic Table).

   Sodium (Na) = 22.997
   Chlorine (Cl) = 35.459

2. Obtain the sum of the atomic weights from the formula for molecular weight.

   Molecular weight of NaCl = 22.997 + 35.459 = 58.456

3. Weigh 58.46 g of NaCl and make up to a total volume of 1000 ml with distilled water.

*To make up 100 ml of a 1 M solution of NaCl:*

1. Same as Step 1 above.

2. Same as Step 2 above.

3. $\dfrac{58.46}{10} = 5.85$.

4. Weigh 5.85 g of NaCl and make up to a total volume of 100 ml with distilled water.

*To make up a 0.1 M (1/10 molar) solution of NaCl:*

1. Same as Step 1 above.

2. Same as Step 2 above.

3. $\dfrac{58.46}{10} = 5.85.$

4. Weigh 5.85 g of NaCl and make up to a total volume of 1000 ml with distilled water.

## 4

# Properties of Some Common Concentrated Acids and Bases

**Table A-4.1**

| Reagent[a] | Molecular weight | Approximate specific gravity of reagent | Molarity of concentrated reagent | Amount of reagent (ml) to make 1 liter of a 1 M solution |
|---|---|---|---|---|
| Hydrochloric acid (HCl) | 36.46 | 1.19 | 12.1 | 82.6 |
| Nitric acid (HNO₃) | 63.02 | 1.42 | 15.8 | 63.3 |
| Hydrofluoric acid (HF) | 20.01 | 1.19 | 29.0 | 34.5 |
| Perchloric acid (HClO₄) | 100.46 | 1.67 | 11.7 | 85.5 |
| Glacial acetic acid (H₃CCO₂H) | 60.05 | 1.05 | 17.4 | 57.5 |
| Sulfuric acid (H₂SO₄) | 98.08 | 1.84 | 18.0 | 55.5 |
| Phosphoric acid (H₃PO₄) | 98.00 | 1.70 | 14.8 | 67.6 |
| Ammonium hydroxide (NH₄OH) | 35.05 | 0.90 | 14.8 | 67.6 |
| Sodium hydroxide (NaOH) | 40.00 | 1.53 | 19.3 | 51.8 |
| Potassium hydroxide (KOH) | 56.11 | 1.46 | 11.7 | 85.5 |

[a] Be sure to look at specifications of the manufacturer of the reagent used in your laboratory, as the molarity and specific gravity vary with the manufacturer.

**5**

# Use of Micropipettors

Many companies manufacture continuously adjustable and fixed pipets to deliver volumes from 1 to 5000 $\mu$l. The micropipettors are generally factory calibrated. However, a careful check of the accuracy and reproducibility of volume delivery should be made to ensure correct usage.

   Micropipettors are available from various companies. The operating directions described below are for the Rainin Pipetman as suggested by the manufacturer. For the micropipettor uscd in your laboratory, be sure to read the instruction manual provided by the manufacturer.

## Operation

1. Set the desired volume by holding the Pipetman body in one hand and turning the volume adjustment knob until the correct volume shows on the digital indicator. The friction O ring of the volume adjustment locks the volume securely. For best volume setting precision, always approach the desired volume by dialing downward (at least one-third revolution) from a larger volume setting.

2. Attach a new disposable tip to the shaft of the pipet. (A major source of error in the use of micropipettors is the variation between sources of disposable tips. It may be desirable to experi-

mentally compare tips from one supplier with those of another.) Press the tip on firmly with a slight twisting motion to ensure a positive, airtight seal.

> Models P-20, P-100, and P-200 use RC-20 yellow tips
> Model P-1000 uses RC-200 blue tips
> Model P-5000 uses C-5000 white tips

3. Depress the plunger to the *first positive stop*. This part of the stroke is the calibrated volume displayed on the digital indicator.

4. Holding the Pipetman vertically, immerse the disposable tip into the sample liquid to a depth of

> 1–2 mm for RC-20 tips (P-20, P-100, P-200)
> 2–4 mm for RC-200 tips (P-1000)
> 3–6 mm for C-5000 tips (P-5000)

5. Allow the push button to return *slowly* to the up position. *Never permit it to snap up.*

6. Wait 1 to 2 seconds to ensure that the full volume of sample is drawn into tip.

7. Withdraw the tip from the sample liquid. Should any liquid remain on the outside of the tip, wipe it carefully with a lint-free cloth, taking care not to touch the tip opening.

8. To *dispense sample,* place the tip end against the side wall of the receiving vessel and depress the plunger slowly to the *first stop*. Wait (longer for viscous solutions) for

> 1 second for RC-20 (yellow) tips
> 1–2 seconds for RC-200 (blue) tips
> 2–3 seconds for C-5000 (white) tips

Then depress the plunger to the *second stop* (bottom of stroke), expelling any residual liquid in the tip.

9. With the plunger fully depressed, carefully withdraw the Pipetman from the vessel, allowing the top to slide along the wall of the vessel.

10. Allow the plunger to return to the *top position.*

11. Discard the tip by depressing the tip ejector button. A fresh tip should be fitted for each sample to prevent carryover between samples.

## Pipetting Guidelines and Precautions

**Consistency** in all aspects of the pipetting procedure will contribute significantly to optimal reproducibility. Therefore attention should be given to

1. *Speed* and *smoothness* during depression and release of the push button

2. *Pressure* on the push button at first stop

3. *Immersion depth*

4. *Minimal angle* from the vertical axis

If an **air bubble** is noted within the tip during intake, dispense the sample into the original vessel, check the tip immersion depth, and pipet more slowly. If an air bubble appears a second time, discard the tip and use a new one.

# 6

# Safe Handling
# of Microorganisms

Many of the procedures used in molecular biology research involve the use of live microorganisms. Whenever such organisms are used, it is essential that laboratory workers adhere rigidly to a microbiology laboratory code of practice and thereby significantly reduce the possibility of causing a laboratory-acquired infection.

The safest way to approach work with live microorganisms is to make the following assumptions:

1. Every microorganism used in the laboratory is potentially hazardous.

2. Every culture fluid contains potentially pathogenic organisms.

3. Every culture fluid contains potentially toxic substances.

The basis of a microbiology laboratory code of practice is that no direct contact should be made with the experimental organisms or culture fluids, e.g., contact with the skin, nose, eyes, or mouth. It must also be noted that a large proportion of laboratory-acquired infections result from the inhalation of infectious aerosols released during laboratory procedures. Below is a list of instructions that forms the basis of a microbiology laboratory code of practice.

1. A laboratory coat that covers the trunk to the neck must be worn at all times.

2. There must be no eating, drinking, or smoking in the laboratory.

3. There must be no licking of gummed labels.

4. Touching the face, eyes, and so on should be avoided.

5. There must be no chewing or biting of pens or pencils.

6. Available bench space must be kept clear, clean, tidy, and free of unessential items such as books and handbags.

7. No materials should be removed from the laboratory without the express permission of the laboratory supervisor or safety officer.

8. All manipulations, such as by pipet or loop, should be performed in a manner likely to prevent the production of an aerosol of the contaminated material.

9. Pipetting by mouth of *any* liquid is strictly forbidden. Pipet fillers, or automatic pipets, are used instead.

10. All manipulations should be performed aseptically, using plugged, sterile pipets, and the contaminated pipets should be immediately sterilized by total immersion in a suitable disinfectant.

11. Contaminated glassware and discarded petri dishes must be placed in lidded receptacles provided for their disposal.

12. All used microscope slides should be placed in receptacles containing disinfectant.

13. It must be recognized that certain procedures or equipment, e.g., agitation of fluids in flasks, produce aerosols of contaminated materials. Lids should be kept on contaminated vessels, when possible.

14. All accidents, including minor cuts, abrasions, and spills of culture fluids and reagents, must be reported to the laboratory supervisor or safety officer.

15. Before leaving the bench, swab your working area with an appropriate disinfectant fluid.

16. Whenever you leave the laboratory, wash your hands with a germicidal soap and dry them with paper towels. Laboratory coats should be removed and stored for future use or laundered. Do not under any circumstances wander into an office or restroom area wearing a potentially contaminated laboratory coat.

# Appendix _____

# 7

# List of Cultures

**Table A-7.1**

| Exercise | Organism[a] | Strain designation | Source |
|---|---|---|---|
| 1 | *Escherichia coli*<br>*Saccharomyces cerevisiae* | — | Any strain from stock culture collection, such as ATCC,[b] or available at your institution |
| 3, 11, 14, 10A | *Saccharomyces cerevisiae* | YNN281 | YGSC[c] |
| 4, 6, 6A, 10A | *Escherichia coli* | LE392(pRY121) | ATCC 37658 |
| 12, 10A | *Escherichia coli* | LE392 | ATCC 33572 |
| 12, 14, 15, 16, 17, 10A | *Saccharomyces cerevisiae* | YNN281(pRY121) | Created by the student in Exercise 11 |
| 1A | *Saccharomyces cerevisiae* | X2180-1A | YGSC or ATCC 26786 |

[a] Genotypes of organisms used are listed in the exercises under Reagents/Supplies.

[b] ATCC, American Type Culture Collection (12301 Parklawn Drive, Rockville, MD 20852); (800) 638-6597; World Wide Web address is http://www.atcc.org/atcc.html.

[c] YGSC, Yeast Genetics Stock Center (Department of Biophysics, University of California, Berkeley, CA 94720); (510) 642-0815; e-mail address is in%"YGSC305@violet.berkeley.edu".

# *8*

# Storage of Cultures and DNA

## Culture Storage

Numerous methods are used to store strains. A permanent, but time-consuming, method for *Escherichia coli* is freeze-drying (lyophilization), by which strains have been kept alive for more than 30 years. Another permanent and useful method for *E. coli* and *Saccharomyces cerevisiae* is ultracold storage at $-70°C$ or in liquid nitrogen. To prevent bursting of the cells during freezing, the culture medium must contain a cryopreservative such as 7% (v/v) dimethyl sulfoxide (DMSO) or 15% (v/v) glycerol. To recover cells from frozen cultures, the stored culture is scraped with a sterile loop or sterile toothpick and ice crystals are allowed to thaw on an agar plate (and then streaked) or placed in sterile growth medium. If the glycerol content is increased to 40–50%, cultures can be stored in a liquid state for several years at $-20°C$.

For convenience, stable strains of *E. coli* can be stored at room temperature as a deep stab culture in airtight tubes containing rich agar. Such stab cultures last for several years. They are prepared by autoclaving a screwcap tube containing liquid agar and allowing the agar to harden. Bacteria are picked with a wire loop from an agar plate or a liquid culture and the loop is stabbed into the rich agar. The tube is then placed in an incubator and the bacteria are grown for 24 hours, after which the tube is sealed with Parafilm and stored at room temperature. The lifetime of a deep stab culture is determined primarily by the quality of the airtight seal, because it is essential that the agar does not dry.

For routine use in the laboratory, a slant culture is used for *E. coli* or *S. cerevisiae*. Liquid agar is placed in a small screwcap tube and sterilized. The tube is tipped to increase the surface area of the liquid and the agar is allowed to harden. The agar surface is then covered with bacteria or yeast by using either a wire loopful of cells or by placing a droplet of culture directly on the surface. The tube is incubated overnight to allow for cell growth and the cap is then tightened. Slant cultures are stored in a refrigerator and will last several months, less if opened frequently.

## DNA Storage

DNA can be conveniently stored in capped and Parafilm-sealed microcentrifuge tubes with the DNA dissolved in sterile water, TE buffer, or 100% ethanol. It is important to remove potentially damaging agents, such as phenol, nucleases, and ethidium bromide, from the DNA before dissolving the DNA in water, buffer, or ethanol. In the short term (several days), storage of DNA in TE buffer, pH 8.0, at 4°C, is acceptable and is preferred if the DNA stock is to be used repeatedly over the course of a few weeks. Freezing and thawing of DNA is an excellent way to introduce nicks in the molecules and should be avoided. In this regard, storing DNA in "frost-free" −20°C freezers should be avoided, as frost–defrost cycles essentially introduce nicks by freeze-thawing. Storage of DNA at −20°C in non-frost-free freezers is acceptable for months. DNA may be stored for longer periods at −70°C.

# Sterilization Methods

**Table A-9.1**

| Method | Time and temperature | Use | Mode of action | Disadvantages |
|---|---|---|---|---|
| Dry heat | 170°C for 1 hour in hot air sterilizer | Glassware, pipets, petri dishes, metals, forceps, etc. | Oxidizes bacterial components | Cannot use for medium sterilization because of high temperatures; most media not stable |
| Intermittent sterilization ("Tyndallization") | 30 minutes a day in flowing steam (100°C) for 3 consecutive days | Media, tissues, liquids, solids | Heat shock material first 30 minutes, and if spores are present, they will germinate and the vegetative cell will then be destroyed in the next cycle | Tedious; time consuming |
| Moist heat | 121°C at 15 psi pressure, 15–30 minutes | Glassware, media, liquids, solids, metals, etc. | Coagulates proteins of microorganisms | Some nutrients are not stable under these conditions, e.g., lactose |
| Filtration | Need Millipore filter, vacuum, Millipore apparatus | Media, liquids | Millipore filter contains pores so small that bacteria cannot pass through | Glassware cannot be sterilized this way |

*(continued)*

**Table A-9.1**   *Continued*

| Method | Time and temperature | Use | Mode of action | Disadvantages |
|---|---|---|---|---|
| Toxic gases (cold sterilization) | Ethylene oxide under slight pressure | Large surface areas, pieces of equipment, hospital rooms, surgery suites, etc. | Kills all life; forms lethal gas | Flammable and sometimes hard to remove all traces |
| Radiation | Ultraviolet (UV) or ionizing ($\gamma$, X-rays) | Media, liquids, solids, pharmaceuticals | Causes genetic mutations, interferes with metabolism and respiration, and leads to death | Does not penetrate glassware easily, requires access to radioactive core or cathode ray tube |

# 10

# Preparation of Stock
# Solutions for Culture Media

**Table A-10.1**

| Component[a] (%, w/v) | Dissolve in: | Chemical type (charge) | To sterilize[b] | Temperature for storage |
|---|---|---|---|---|
| Amino acids | | | | |
| Alanine (0.5) | $H_2O$[c] | Aliphatic (0) | Autoclave | 4°C |
| Aspargine (0.5) | $H_2O$ | Amide (0) | Autoclave | 4°C |
| Arginine (1) | $H_2O$ | Basic (+) | Autoclave | 4°C |
| Aspartic acid (1) | 0.2 $N$ NaOH | Acidic (−) | Filter | Room temperature |
| Cysteine (0.5) | $H_2O$ | Sulfur (0) | Filter | 4°C |
| Glutamic acid (1) | 0.2 $N$ NaOH | Acidic (−) | Filter | Room temperature |
| Glycine (2) | $H_2O$ | Aliphatic (0) | Autoclave | 4°C |
| Isoleucine (1) | $H_2O$ | Aliphatic (0) | Autoclave | 4°C |
| Leucine (1) | $H_2O$ | Aliphatic (0) | Autoclave | Room temperature |
| Lysine (1) | $H_2O$ | Basic (+) | Autoclave | 4°C |
| Methionine (1) | $H_2O$ | Sulfur (0) | Autoclave | 4°C |
| Threonine (4) | $H_2O$ | Hydroxyl (0) | Autoclave | Room temperature |
| Tyrosine (0.2) | 0.2 $N$ NaOH | Aromatic (0) | Filter | Room temperature (dark) |
| Tryptophan (1) | 0.1 $N$ HCl | Aromatic (0) | Filter | 4°C (dark) |

(*continued*)

**Table A-10.1**  *Continued*

| Component[a] (%, w/v) | Dissolve in: | Chemical type (charge) | To sterilize[b] | Temperature for storage |
|---|---|---|---|---|
| Serine (8) | $H_2O$ | Hydroxyl (0) | Autoclave | 4°C |
| Phenylalinine (1) | $H_2O$ | Aromatic (0) | Autoclave | Room temperature (dark) |
| Proline (2) | $H_2O$ | Imino (0) | Autoclave | 4°C |
| Valine (3) | $H_2O$ | Aliphatic (0) | Autoclave | 4°C |
| Histidine (1) | $H_2O$ | Basic (+) | Autoclave | 4°C |
| Purines and pyrimidines | | | | |
| Adenine sulfate (0.2) | 0.1 *N* HCl | Purine | Filter | Room temperature |
| Guanine (0.5) | 0.1 *N* HCl | Purine | Filter | 4°C |
| Thymine (0.1) | $H_2O$ | Pyrimidine | Autclave | 4°C |
| Cytosine (0.5) | $H_2O$ | Pyrimidine | Autoclave | 4°C |
| Uracil (0.2) | $H_2O$ | Pyrimidine | Autoclave | Room temperature |

[a] All amino acids are in the L form.
[b] For filtration, use membrane filter with a 0.45- or 0.2-$\mu$m pore size.
[c] Use distilled water for preparation of all stock solutions.

# *11*

# Growth in Liquid Medium

## Growth Curve

Cells growing in a batch (shake flask) culture generally experience four distinct growth phases: a lag phase, an exponential (log) phase, a stationary phase, and a death phase. Cells in the lag phase, in which an aliquot of cells from an older culture has been transferred into fresh medium, do not grow right away. These cells must adjust to the new medium before growth will begin at a rapid rate. The length of the lag phase is dependent on a number of factors: age and genotype of the inoculum, temperature, nutrient levels of both the old and new media, aeration, and the concentration of toxins that may have been formed in the old medium. The lag phase of *Saccharomyces cerevisiae* at 30°C in YEPD lasts for about 3 hours. For *Escherichia coli* at 37°C, the lag phase is usually 10–60 minutes.

Once the cells begin growing rapidly, they are said to enter the exponential, or logarithmic (log), growth phase. Cells in this phase are growing rapidly and, unlike cells in the lag and stationary phases, most cells are in the same physiological state. The growth rate during log phase is dependent on the nutrient level and aeration of the medium. Oxygen is frequently the limiting factor in yeast cultures, and the cultures must be shaken rapidly for sufficient $O_2$ to dissolve. The doubling time of *S. cerevisiae* grown in YEPD at 30°C may be between 90 and 100 minutes, while a 200-minute doubling time may be expected in minimal medium. The presence of an introduced plasmid may slow down growth since plasmid replication requires energy and metabolites. At 37°C, a typical *E. coli* strain doubles in 20–30 minutes in rich medium and in 50–60 minutes in minimal medium. The growth rate constant ($\mu$) serves to define the rate of growth of a culture during balanced growth: $\mu = \ln 2 \div g$ (mean doubling time or generation time). For further discussion of this concept and derivation of this formula, see Mandelstam *et al.* (1982).

As nutrients within the flask are consumed and inhibitory products accumulate, the growth rate slows and eventually stops as the culture enters the stationary phase. Cells in a stationary phase culture are not all in the same physiological state; some are dividing while others are dying. Only the overall population size remains constant. As more nutrients are depleted, more cells die than are produced, and the culture enters the death phase.

## Aeration

*Escherichia coli* and *Saccharomyces cerevisiae* can grow both aerobically or anaerobically, although anaerobic growth does not occur with all carbon sources. With glucose, anaerobic growth occurs, but at about 10% of the rate of aerobic growth. At a density of approximately $10^7$ cells/ml in liquid medium, oxygen cannot diffuse from the atmosphere to the cells fast enough for aerobic growth, and unless the medium is shaken or directly aerated by bubbling, growth of a culture slows considerably at this cell density. Even with aeration, the oxygen supply is inadequate above a cell concentration of $2$–$3 \times 10^9$ cells/ml (*E. coli*) or $2$–$3 \times 10^8$ cells (*S. cerevisiae*).

## Inoculation and Subculture

Since many cells in an inoculum obtained from a slant culture are dead, most experiments begin by inoculating a small volume of growth medium and then growing the cells overnight. In the morning, cell growth will have stopped and the cell density will usually be

2–3 × 10⁹ cells/ml for *E. coli* in rich medium and 2–4 × 10⁸ cells/ml for *S. cerevisiae* in rich medium. To obtain actively growing cells for any particular experiment, the overnight culture is diluted 10–100 times and regrown for several hours, at which time the cells are usually in exponential growth. Also, if an exponentially growing culture is rapidly chilled to below 8°C (by shaking a growth flask in ice water), stored at 4°C, and then rewarmed rapidly to the original temperature at which the culture had been growing, growth will resume without a lag.

## Reference

Mandelstam, J., McQuillen, K., and Dawes, I. (1982). "Biochemistry of Bacterial Growth," 3rd Ed. Halsted Press, Oxford, UK.

*Appendix* _____

# *12*

# Determination of Viable Cells

One cell can multiply until a single visible colony forms. This is the basis of counting cells by plating, because a count of the number of colonies produced by a particular volume of a culture indicates the number of viable cells in the culture. In either an exponentially growing culture or a culture that is in early stationary phase, usually all cells can form colonies. Thus the colony count approximately equals the number of viable cells.

To obtain a reliable count by plating, it is essential that single colonies are not formed by two or more cells. Accordingly, the number of cells placed on an agar surface should not exceed a few hundred and the cells should be spread evenly. Since the cell concentration used in laboratory experiments is usually more than $10^6$ cells/ml, the culture must be diluted prior to plating. The standard procedure is to make a series of sequential 10- to 100-fold dilutions until the cell concentration is a few thousand cells per milliliter. Then, a 0.1-ml sample is spread on an agar surface; this operation (dilution and spreading) is called plating. To avoid error that might arise by transferring very small volumes, a volume of 0.05–0.5 ml is usually transferred. We will use 10- and 100-fold dilutions prepared, respectively, by adding 0.5 ml of cells to 4.5 ml of sterile diluent and 0.05 ml of cells to 4.95 ml of sterile diluent. (Many laboratories use 1 to 9.0 and 0.1 to 9.9.) Each dilution must be done with a separate pipet; otherwise cells remaining from an earlier dilution may be carried over to a later dilution tube. A major cause of dilution errors is the transfer of liquid on the outside of the pipet. This could be avoided by wiping the pipet, but is usually not possible because of resulting contamination. Therefore, to minimize errors, it is best to submerge the tip as little below the surface of the liquid as possible, touch the side of the tube to remove any adhering droplets, and then blow out the liquid into the diluent. The necessity to blow out the liquid is the reason that serological (blowout) pipets are used rather than analytical (to deliver) pipets.

When the final aliquot of 0.1 ml of diluted cells is placed on the agar, the droplet is spread over the surface with a glass spreader (hockey stick). The spreader is sterilized by dipping it in ethanol, shaking off the excess liquid, and igniting the remaining alcohol in the flame from a Bunsen burner. The spreader is cooled by touching the agar surface and then used to spread the droplet uniformly over the surface. If liquid remains on the agar, cells will drift through the liquid and, after cell division, two colonies might form from one cell initially deposited. Plates prepared 1 day in advance usually absorb the liquid rapidly; absorption is accelerated by spreading the droplet as thinly as possible. After the surface of the plate is dry, the plate is placed in an incubator. As the plate warms, liquid is exuded from the uppermost surface of the agar. Thus, it is advisable to invert the plates in the incubator to avoid puddles on the surface of the agar because cells will drift through the liquid. This also prevents droplets, which might condense on the inside surface of the top plate, from falling on the agar.

# *13*

# Determination of Cell Mass

Growth of cultures can be monitored by spectrophotometry. The number of photons scattered is proportional to the mass of the cells in a sample (except for very concentrated cultures, as discussed in the next paragraph) or, for particular growth conditions, to the cell concentration. However, the geometry of each spectrophotometer determines how much scattered light falls onto a photodetector. Therefore, a calibration curve that relates cell concentration and absorbance must be made for each spectrophotometer. This is done by measuring the absorbance of suspensions of cells at different concentrations and determining the cell concentration of each suspension by direct counting or by plating on agar and measuring the number of colonies formed.

It is important to note that the absorbance is a measure of cell mass rather than cell number. Cell size varies with growth phase, so it is best to calibrate the spectrophotometer with exponentially growing cells because such cells are used in most experiments. Cell size also varies from one growth medium to the next, decreasing as the medium becomes poorer, so a calibration curve is needed for each growth medium, and often for each strain. As a rough guide, for many *Escherichia coli* strains, $A_{600} = 1.0$ for $\sim 8 \times 10^8$ cells/ml and for *Saccharomyces cerevisiae*, $A_{600} = 1.0$ for $\sim 3 \times 10^7$ cells/ml.

If the cell density is too high, a photon may be deflected away from the photodetector by one cell and then back again by a second cell. This effect causes the absorbance to be lower than if multiple scattering were not occurring; it becomes important at values of $A_{600}$ above 0.7. Thus, when the concentration of a dense culture is to be determined by spectrophotometry, the culture is diluted prior to reading. The measured value is then corrected by the dilution factor.

Wavelengths other than 600 nm can be employed in determining cell density and, in fact, the sensitivity increases as the wavelength decreases. Wavelengths as low as 400 nm may be used, but not with all rich media. These media usually absorb short-wavelength light significantly and the absorption complicates the measurements.

# *14*

# Determination of Cell Number

To find the total number of cells in a culture, direct microscopic counting is the method of choice. For yeast cells, a hemocytometer can be used. Magnification of ×100 to ×400 is needed to visualize the counting chamber and the suspended cells. For bacterial cells, a Petroff–Hauser chamber and ×1000 magnification is required because of the small size of the cells. Direct counting is often advantageous because of the opportunity afforded to visualize directly the morphology of the cell studied. Aberrant or unusual morphology could indicate suboptimal growth conditions or the presence of an altered cell genotype or phenotype.

## Charging the Chamber

Vortex a cell suspension. Withdraw some of the suspension into a Pasteur pipet. Deposit a small drop on the polished surface of the counting chamber next to the edge of the cover glass. The suspension will enter the chamber by capillary action. A properly filled chamber has cells filling only the space between the cover glass and counting chamber. No fluid should run down into the moat.

## Performing the Count

Place the hemocytometer on the stage of a compound microscope and focus on the grid lines under low power. For yeast counting, use the middle square (see Figure A-14.1; the middle square is encircled and contains 25 squares, each bounded by 3 grid lines indicated by heavy black lines. Each of those 25 squares contains 16 squares). Adjust the light intensity so that both cells and lines are plainly visible. Depending on the cell numbers, either count all the cells in the middle square or choose five of the squares indicated by an "X" mark within this middle square (Figure A-14.1). Include those cells that touch the boundary lines but do not overlap the boundaries into other areas.

## Calculations

The middle squares has a volume of 0.1 mm$^3$. The number of cells in the middle square multiplied by 10 will yield the number of cells in 1.0 mm$^3$ of the culture. That number of cells multiplied by 1000 will yield the number of cells in 1.0 cm$^3$ or in 1.0 milliliter (ml) of the culture. Therefore, if the total middle square is counted, multiply that number by 10$^4$ to yield the cells per milliliter. Alternatively, if five squares within the middle square are counted, that cell number should be multiplied by $5 \times 10^4$ to yield the cells per milliliter. Whichever volume is counted, a total of more than 100 cells should be counted to obtain an accurate cell count.

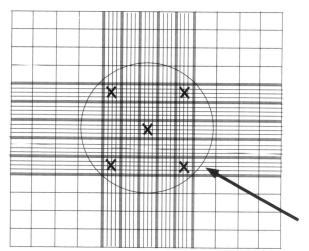

**Figure A-14.1**  Hemocytometer chamber. The middle, encircled area is seen under 100-fold magnification.

# *15*

# Nomenclature of Strains

## Bacteria

The standard nomenclature for genes and genetic markers for bacteria is that of Demerec *et al.* (1966). A three-letter italicized or underlined symbol is used to describe a genetic locus, for example, *his* (histidine) or *gal* (galactose). All histidine-requiring mutants are written *his* or *his⁻*. Often there are several enzymes in a locus; for example, several enzymes are needed to synthesize histidine; a deficiency of any of these will be *his⁻*. A capital letter following *his* is used to distinguish the loci: for example, *hisA, hisB,* and so on. Usually, each locus represents a particular polypeptide chain or regulatory element (operator or promoter) in the system. Each independently isolated mutation within a particular locus is given a number, for example, *hisA38*. Not all *hisA* mutations are identical or yield exactly the same phenotype but all *hisA38* mutations are identical. The phenotype of a cell is written with a capital letter and is not italicized; a haploid cell carrying the *hisA38* marker has the His⁻ phenotype, whereas a wild-type, histidine prototroph is designated His⁺.

The genotype of a strain is usually given by listing all loci known to be different from the wild type. When necessary, a (+) is appended to a genetic symbol to indicate a wild-type allele of a locus; for example, *hisC⁺* means a wild-type gene.

A detailed linkage map of *Escherichia coli* has been published by Bachmann (1990). In addition to the map, the article by Bachmann is a valuable source of references to original research papers describing mutations found in many strains; it also contains useful references to techniques used in modern microbial genetics. Most strains of *E. coli* can be obtained free of charge from B. Bachmann (Department of Human Genetics, Yale University School of Medicine, 333 Cedar Street, New Haven, CT 06510) (see also Appendix 21 for World Wide Web address).

## *Saccharomyces cerevisiae*

For *Saccharomyces cerevisiae,* the following are examples used in genetic nomenclature:

| | |
|---|---|
| *URA3* | Locus or dominant allele |
| *ura3* | Locus or recessive allele that results in the requirement of uracil |
| *URA3*[+] | Wild-type allele |
| *ura3-52* | Specific allele or mutation at the *URA3* locus |
| Ura[+] | Phenotypic designation of a strain not requiring uracil |
| Ura[−] | Phenotypic designation of a strain requiring uracil |
| Ura3p | Protein encoded by the *URA3* gene |

The gene symbols differ somewhat from those for bacteria in that the genetic locus is identified by a number rather than a letter following the three-letter gene symbol. Recessive alleles are symbolized by lowercase, italicized letters, whereas dominant alleles are represented by uppercase, italicized letters. Different alleles for the same locus are differentiated by a number separated by a hyphen and following the locus number. There are a number of exceptions to these general rules. For example, the wild-type allele of the mating type locus are designed *Mat*a and *Mat*α, and mutations are *mat*a-*1* and *mat*α-*1*. For further discussion of genetic nomenclature, see Sherman (1981) and Kaiser *et al.* (1994).

A detailed linkage of *S. cerevisiae* has been published as well (Mortimer and Schild, 1985). Strains of *S. cerevisiae* can be purchased for a minimal charge from the Yeast Genetics Stock Center (see

Appendix 7). The American Type Culture Collection (see Appendix 7 and Appendix 21 for World Wide Web address) is also an excellent source for *E. coli* and *S. cerevisiae* strains.

## References

Bachmann, B. (1990). Linkage map of *Escherichia coli* K-12, ed. 8. *Microbiol. Rev.* **54,** 130–197.

Demerec, M., Adelberg, E. A., Clark, A. J., and Hartman, P. E. (1966). A proposal for a uniform nomenclature in bacterial genetics. *Genetics* **54,** 61–76.

Kaiser, C., Michaelis, S., and Mitchell, A. (1994). "Methods in Yeast Genetics." Cold Spring Harbor Laboratory, Cold Spring Harbor, New York.

Mortimer, R. K., and Schild, D. (1985). Genetic map of *Saccharomyces cerevisiae*, ed. 9. *Microbiol. Rev.* **49,** 181–213.

Sherman, F. (1981). Genetic nomenclature. In "Molecular Biology of the Yeast *Saccharomyces*" (Strathern, J. N., *et al.*, eds.), pp. 639–640. Cold Spring Harbor Laboratory, Cold Spring Harbor, New York.

# 16

# Glassware and Plasticware

Experiments in molecular biology require that all items of glassware and other equipment be scrupulously clean before use. This is particularly important for reactions involving proteins and nucleic acids carried out in small volumes, because dirty test tubes, bacterial contamination, or traces of detergent can easily inhibit reactions or accidentially degrade nucleic acids by the action of unwanted nucleases. Following thorough washing (dichromate–sulfuric acid solution is very effective for cleaning glass pipets), glass or plastic vessels or pipets should be thoroughly rinsed in distilled or double-distilled water and sterilized by autoclaving. Alternatively, heat-resistant glassware can be baked in an oven at 150°C for 1 hour.

Glassware can be treated with 0.1% (v/v) diethyl pyrocarbonate overnight at 37°C as a precaution against ribonuclease contamination. The diethyl pyrocarbonate is removed by heating the drained glassware at 100°C for 15 minutes. This procedure is not normally required for sterile disposable plasticware.

Very small quantities of nucleic acids can be absorbed onto the surface of glassware. The problem is significantly reduced if small polypropylene (e.g., microcentrifuge or Eppendorf) test tubes are used, but if these are not available, you may wish to coat glassware with silicone. To do this, soak or rinse the glassware in a 5% (v/v) solution of dichlorodimethylsilane in chloroform. This solution can be stored at room temperature in a fume hood virtually indefinitely and used as required. Following this silicone treatment, thoroughly rinse the glassware and bake at 180°C overnight.

Many of the experiments with nucleic acids involve the use of solutions containing phenol and/or chloroform. These solutions can be successfully used in glass or polypropylene (Eppendorf) tubes, but attack many other forms of plastic, particularly polycarbonate, as used in many types of centrifuge tubes. Check the manufacturer recommendations if you are doubtful about using a particular type of plastic tube and remember that sealing films are also soluble in phenol and chloroform.

# *17*

# Preparation of Tris and EDTA

## 1 *M* Tris Stock

Weigh out 121.1 g of Tris base [tris(hydroxymethyl)aminomethane, MW 121.1 g/mol] and place it in a 2000-ml beaker containing a magnetic stir bar. Add 800 ml of distilled $H_2O$ and dissolve the Tris, using a magnetic stirrer. Adjust the pH to the desired value by addition of concentrated HCl. The addition of HCl should be done carefully, while the solution is being stirred and monitored for pH, to avoid a drastic drop in the pH. The amount of concentrated HCl to be added differs according to the desired pH of the Tris solution (Table A-17.1).

**Table A-17.1   pH of Tris Stock Solution**

| pH desired | Volume of HCl (ml) |
|:---:|:---:|
| 7.4 | ~70 |
| 7.6 | ~60 |
| 7.8 | ~50 |
| 8.0 | ~42 |

Check the pH, pour the solution into a graduated cylinder (or volumetric flask), and bring the volume of the solution up to 1000 ml using distilled $H_2O$. Mix and dispense as necessary and sterilize by autoclaving. Tris stocks of different molarities are prepared in a similar manner.

## Notes

1. It is desirable to obtain a calomel electrode, which is an electrode compatible with Tris. These can be purchased from most manufacturers. With silver pH electrodes, Tris precipitates, causing clogging of the electrode tip.
2. The pH of a Tris buffer varies greatly with temperature. For example, a pH of 8.0 at 25°C becomes 8.58 at 5°C and 7.71 at 37°C. For experiments in this laboratory book, the pH of Tris at 25°C (room temperature) is used even if the solutions are cooled or heated. However, for some experiments, the pH variation with temperature must be taken into account.
3. A method for preparation of 0.05 $M$ Tris stock solution if Tris base and Tris-HCl are available involves simply mixing these two compounds as indicated in Table A-17.2.

**Table A-17.2  Preparation of 50 m$M$ Tris Buffer**

| pH at 25°C | Tris-HCl (g/liter) | Tris base (g/liter) |
|------------|--------------------|---------------------|
| 7.4 | 6.61 | 0.97 |
| 7.6 | 6.06 | 1.39 |
| 7.8 | 5.32 | 1.97 |
| 8.0 | 4.44 | 2.65 |

## 0.5 $M$ EDTA Stock

Weigh out 186.1 g of disodium EDTA (disodium ethylenediaminetetracetic acid dihydrate, MW 372.2 g/mol) and place it in a 2000-ml beaker. Add 800 ml of distilled $H_2O$ and stir vigorously, using a stir bar, on a magnetic stirrer. Adjust the pH to 8.0 using NaOH (concentrated) to facilitate the dissolution of the EDTA. This can be accomplished by addition of concentrated NaOH liquid, using a Pasteur pipet, or by addition of ~20 g of NaOH pellets while mixing and monitoring the pH of the EDTA solution. This is then brought up to a volume of 1000 ml in a graduated cylinder (or volumetric flask), mixed, and dispensed into aliquots for sterilization by autoclaving.

# 18

# Basic Rules for Handling Enzymes

**Note**

This is a reprint of the Boehringer Mannheim Biochemicals primer, which was first published in the December 1985 issue of *BM Biochemica*.

*For the novice:* Basic hints to guide you through your first enzymatic reaction.

*For the expert:* A refresher course and an aid for training your students.

1. For best stability, enzymes should be stored in their original commercial form (lyophilized, ammonium sulfate suspension, etc.), undiluted, and at the appropriate temperature as specified on the label.

2. For enzyme solutions and assay buffers, use the highest purity $H_2O$ available. Glass-distilled $H_2O$ is best. Deionized $H_2O$, especially if passed through an old deionizing filter or a reverse osmosis device, may contain traces of organic contaminants that inhibit enzymes.

3. Enzymes should be handled in the cold (0–4°C). Dilute for use with ice-cold buffer or distilled water, as appropriate for each

enzyme. While using the enzyme solution or suspension at the bench, keep it in an ice bath or ice bucket.

4.  Dilute enzyme solutions are generally unstable. The amount of enzyme required for the experiment should be diluted within 1–2 hours of use. Enzymes should not be diluted for long-term storage. Enzymes, especially those that have been diluted, should be checked for activity periodically to ensure that any slight loss in activity is taken into account when designing an experimental protocol.

5.  Do not shake a crystalline suspension (e.g., ammonium sulfate suspension) because oxygen tends to denature the enzyme. The material should be resuspended with gentle swirling or by rolling the bottle on the laboratory bench. Once the enzyme crystals have been uniformly resuspended, remove the amount needed with a pipet. In many cases, the enzyme crystals may be used directly in assay procedures.

6.  Do not freeze crystalline suspensions. Freezing and thawing in the presence of high salt causes denaturation and loss of activity.

7.  Vials containing lyophilized enzymes (as well as cofactors, such as NADH and NADPH) should be warmed to room temperature before opening. This prevents condensation of moisture onto the powder, which can cause loss of activity or degradation. If the reagent is hygroscopic, one such mishandling may well ruin the entire vial.

8.  Avoid repeated freeze-thawing of dilute enzymes and lyophilizates in solution. Store in small aliquots. Thaw one portion at a time and store that portion, once thawed, at 4°C. The stability of individual enzymes may vary greatly and often should be determined empirically under your exact conditions.

9.  Detergents and preservatives should be used with caution, because they may affect enzyme activity. Sodium azide, for example, inhibits many enzymes that contain heme groups (e.g., peroxidase). Detergents added at concentrations above their critical

micellar concentration form micelles, which may entrap and denature the enzyme.

10. Enzymes should be handled carefully to avoid contamination of any kind. Use a fresh pipet for each aliquot that is removed from the parent vial. Never return unused materials to the parent vial. Wear gloves to prevent contaminating the enzyme with proteases, DNases, RNases, and inhibitors often found on fingertips. Never pipet by mouth.

11. Adjust the pH of the enzyme buffer at the temperature at which it will be used. Many common buffers (Tris, glycylglycine, BES, ACES, TES, Bicine, and HEPES) change pH rapidly as the temperature changes. For instance, Tris buffer decreases 0.3 pH units for every 10°C rise in temperature. A solution of Tris, adjusted to pH 7.5 at 25°C, will have a pH of 8.1 at 4°C or 7.2 at 37°C. The changes in pH per 10°C temperature change for other buffers are as follows: ACES, −0.20; BES, −0.16; Bicine, −0.18; glycylglycine, −0.28; HEPES, −0.14; TES, −0.20 (Good, *et al.*, 1966).

12. The absorbance at 280 nm, widely used to determine quickly the protein concentration of an enzyme solution, actually is due to the presence of tyrosine and tryptophan in the protein. If an enzyme (e.g., superoxide dismutase) contains a low amount of these two amino acids, it will not absorb significantly at 280 nm.

## References

Good, N. E., Winget, G. D., Winter, W., Connolly, T. N., Izawa, S., and Singh, R. M. M. (1966). Hydrogen ion buffers for biological research. *Biochemistry* **5**, 467–477.

Detailed information is available on many enzymes. The most complete references include the following:

*Methods in Enzymology,* published by Academic Press; Editors-in-Chief: John N. Abelson and Melvin I. Simon. There are more than 200 volumes in this series, covering an extensive range of topics.

*The Enzymes,* 3rd Ed., published by Academic Press; edited by Paul D. Boyer. This excellent, broad series focuses more on the physical and biological properties of enzymes and less on methodology than does *Methods in Enzymology.*

*Methods of Enzymatic Analysis,* 3rd Ed., published by Verlag Chemie; Editor-in-Chief: Hans U. Bergmeyer. This book provides an in-depth discussion of analytical techniques that use enzymes or that assay enzymes.

# *19*

# Effects of Common Contaminants on Protein Assays

Protein purification schemes employ a variety of chemical reagents that facilitate separation procedures. Residual levels of these reagents may prevent accurate assessment of protein quantity. Similarly, cellular contaminants in protein preparations are common sources of error in protein assays. The manner and extent to which assays are affected by interfering substances are dependent on which detection methods and biochemicals are utilized. Therefore, choice of a protein assay that will be minimally affected by the experimental system is critical to successful quantitation of protein preparations.

This book demonstrates the two most commonly practiced protein assays: the Bradford assay and the Lowry assay (Exercise 13). Table A-19.1 summarizes the effect specific substances may have on the accuracy of each of these methods for measuring protein. A substance is designated ($-$) for no interference, ($+$) for clearly detectable interference, and ($\pm$) for having the potential of interfering.

**Table A-19.1  Effects of Common Contaminants on Protein Assays[a]**

| Interfering substance | Bradford assay | Lowry assay |
|---|---|---|
| Ammonium sulfate | − | − |
| Cesium chloride | + | + |
| Citrate | − | + |
| Dextran sulfate | + | ± |
| DNA | − | ± |
| EDTA | − | + |
| Ethanol | − | − |
| Glucose | − | ± |
| Glycerol | − | + |
| Glycine | − | + |
| Heparin | + | − |
| HEPES | − | + |
| Magnesium | − | + |
| Manganese chloride | − | + |
| Mercaptoethanol | − | + |
| MOPS | − | + |
| Nonidet P-40 | + | + |
| Penicillin | − | + |
| Percoll | + | + |
| Polyacrylamide | − | + |
| Potassium chloride | − | + |
| RNA | − | ± |
| Sodium chloride | − | ± |
| SDS | ± | ± |
| Sodium phosphate | − | ± |
| Sucrose | ± | + |
| Toluene | + | ± |
| Tris | − | + |
| Triton X-100 | ± | + |
| Urea | − | − |

[a] Derived from "Protein Assays," *American Biotechnology Laboratory,* July 1988, pp. 29–37.

# *20*

# Manufacturers' and Distributors' Addresses

This appendix is designed to assist the instructor in obtaining the reagents, equipment, and supplies needed for teaching this course. For a complete listing of biotechnology products and instruments, see *Buyer's Guide Edition*, published by *American Laboratory,* or *The Lab Manual Sourcebook,* published by Cold Spring Harbor Laboratory Press. Access to manufacturers and their products is also available for many companies through the Internet (see Appendix 21).

## Major Distributors of General Laboratory Equipment and Supplies

Baxter Scientific Products (General Offices)
1430 Waukegan Road
McGaw Park, Illinois 60085
(312) 689-8410

Beckman Instruments, Inc.
Spinco Division
1117 California Avenue
Palo Alto, California 94340
(415) 857-1150

B. Braun Biotech
999 Postal Road
Allentown, Pennsylvania 18103
(215) 266-6262

Du Pont (Sorvall)
Biotechnology Systems
Wilmington, Delaware 19818
(800) 551-2121

Fisher Scientific (varies with location)
Headquarters:
711 Forbes Avenue
Pittsburgh, Pennsylvania 15219
(412) 562-8300

Millipore Corporation
80 Ashby Road
Bedford, Massachusetts 01730
(800) 645-5476

PGC Scientifics
9161 Industrial Court
Gaithersburg, Maryland 20877
(301) 840-1111

Schleicher & Schuell, Inc.
P.O. Box 2012
Keene, New Hampshire 03431
(800) 245-4024

Thomas Scientific
99 High Hill Road
P.O. Box 99
Swedesboro, New Jersey 08085
(800) 524-0027

# Chromatography

Bio-Rad Laboratories
1414 Harbour Way South
Richmond, California 94804
(800) 227-3259 West
(800) 645-3227 East

Pharmacia LKB Biotechnology, Inc.
800 Centennial Avenue
Piscataway, New Jersey 08854
(800) 558-7110

5 Prime → 3 Prime, Inc.
5603 Arapahoe Road
Boulder, Colorado 80303
(800) 533-5703

Qiagen
9600 De Soto Avenue
Chatsworth, California 91311
(800) 426-8157

Supelco, Inc.
P.O. Box B
Bellefonte, Pennsylvania 16823
(814) 359-3441

# Electrophoresis Apparatus and Power Supplies

Bio-Rad Laboratories, *see* Chromatography

Hoefer Scientific Instruments
654 Minnesota Street
P.O. Box 77387
San Francisco, California 94107
(800) 227-4750

Idea Scientific Company
P.O. Box 2078
Corvallis, Oregon 97339
(503) 758-0999

## Fine Chemicals

Aldrich Chemical Company, Inc.
940 West Saint Paul Avenue
Milwaukee, Wisconsin 53223
(800) 558-9160
(414) 273-3850

Amersham Corporation
2636 South Clearbrook Drive
Arlington Heights, Illinois 60005
(800) 323-9750

Ameresco, Inc.
30175 Solon Industrial Parkway
Solon, Ohio 44139
(800) 366-1313

Calbiochem Corporation
P.O. Box 12087
San Diego, California 92112
(800) 854-3417

FMC Bioproducts
5 Maple Street
Rockland, Maine 04840
(800) 341-1475
(207) 594-3200

J.T. Baker Chemical Company
22 Red School Lane
Phillipsburg, New Jersey 08865
(201) 859-5411

Sigma Chemical Company
P.O. Box 14508
St. Louis, Missouri 63178
(800) 325-3010

## Microbiological Media and Supplies

BBL Microbiology Systems
P.O. Box 243
Becton Dickinson Co.
Cockeysville, Maryland 21030
(800) 638-6663

Difco Laboratories
P.O. Box 1058
Detroit, Michigan 48232
(313) 961-0800

ReplicaTech, Inc.
P.O. Box 7484
Princeton, New Jersey 08543
(800) 748-7487

See also Major Distributors of General Laboratory Equipment and
Supplies, listed above

## Microbiological Strains

American Type Culture Collection (ATCC)
12301 Park Lawn Drive
Rockville, Maryland 20852
(800) 638-6597

Yeast Genetics Stock Center
Department of Biophysics
University of California
Berkeley, California 94720
(510) 642-0815

## Oligonucleotides

Bioserve Biotechnologies
1050 West Street
Laurel, Maryland 20707
(301) 470-3362

Genosys Biotechnologies Inc.
1442 Lake Front Circle
The Woodlands, Texas 77380
(800) 234-5362

Integrated DNA Technologies, Inc.
1710 Commercial Park
Coralville, Iowa 52241
(800) 328-2661

Midland Certified Reagent Co.
3112-A West Cuthbert Avenue
Midland, Texas 79701
(800) 247-8766

## PCR Reagents and Apparatuses

Ericomp, Inc.
6044 Cornerstone Court
San Diego, California 92121
(800) 541-8471

MJ Research
149 Grove Street
Watertown, Massachusetts 02127
(800) 729-2165

Perkin-Elmer
761 Main Avenue
Norwalk, Connecticut 06859
(800) 762-4000

Promega Corporation
2800 South Fish Hatchery Road
Madison, Wisconsin 53711
(800) 356-9526

## Restriction Enzymes and Reagents for Molecular Biology

Boehringer Mannheim Biochemicals (BMB)
9115 Hague Road
P.O. Box 50816
Indianapolis, Indiana 46250
(317) 576-2771

Clontech Laboratories Inc.
4030 Fabian Way
Palo Alto, California 94303

Gentra Systems, Inc.
P.O. Box 13159
Research Triangle Park, North Carolina 27709
(800) 866-3039

Kodak Scientific Imaging Systems
25 Science Park
New Haven, Connecticut 06511
(800) 225-5352

Life Technologies Inc. (GIBCO/BRL)
P.O. Box 6009
Gaithersburg, Maryland 20877
(800) 638-8992

New England Biolabs, Inc. (NEB)
32 Tozer Road
Beverly, Massachusetts 01915
(800) 632-5227
(617) 927-5054

Promega Corporation, *see* PCR Reagents

Sigma, *see* Fine Chemicals

Stratagene Cloning Systems
11099 North Torrey Pines Road
La Jolla, California 92037
(800) 548-1113

United States Biochemical Corporation
P.O. Box 22400
Cleveland, Ohio 44122
(800) 321-9322

# 21

# Surfing the Bionet: World Wide Web Addresses

A revolution is taking place in biology with the help of the Internet (Net) and the World Wide Web (WWW). Information of immense value is now accessible electronically through the WWW, which merges databases from around the world. These information resources are crucial at every level of biology, providing a comprehensive and coherent view of the state of the art to students and researchers.

The list of World Wide Web sites (Table A-21.1) is intended to aid access to various databases. Although WWW sites are constantly evolving and expanding, this list should help you get started using the Net. It is easy to connect your computer to the WWW using Mosaic, Netscape, or other readily available software. We trust that "surfing the bionet" will become an instructive and interesting pasttime of students and instructors in this course.

**Table A-21.1    Addresses on World Wide Web: Electronically Accessible Data of Interest to Biotechnologists and Molecular Biologists**

| Address[a] | Subject |
|---|---|
| webcrawler.cs.washington.edu/ webcrawler/webquery.html | Search the WWW by using this address and typing in any keyword |
| golgi.harvard.edu/biopages/all.html | WWW virtual library on the biosciences (e.g., mycology, biotechnology, microbiology, genetics), biodiversity, molecular biology, and biochemistry and biophysics, universities, etc. |
| www.public.iastate.edu/~pedro/ research_tools.html | WWW virtual library; collection of links to information and services useful to molecular biologists |
| www.ncbi.nlm.nih.gov | National Center for Biotechnology Information; a genetic sequences database of interest to molecular biologists |
| macserver.molbio.gla.ac.uk | Dictionary of cell biology |
| best.gdb.org/best.html | Identify and locate researchers with interests similar to your own |
| kumchttp.mc.ukans.edu/research/ fgsc/main.html | Fungal genetics (primarily *Neurospora, Aspergillus,* and related fungi and certain cloned genes from these fungi) |
| cgsc.biology.yale.edu/top.html | Database of *Escherichia coli* genetic information including strains, genes, etc. |
| www.atcc.org/atcc.html | American Type Culture Collection; catalog of strains, plasmids, etc. |
| genome-www.stanford.edu | *Saccharomyces cerevisiae* database including list of strains and genes, sequence of genome, etc. |
| alpha.genebee.msu.su:80/supplier or web.frontier.net/MEDMarket/ indexes/indexmfr.html | Lists of suppliers of equipment, chemicals, and reagents needed for biological research; also information on biotech companies |
| www.atcg.com/atcg | Useful information on restriction endonucleases |

[a] Precede all addresses with **http://**

# Glossary

These terms are defined in the context of this course. Commonly used jargon is also defined.

**absorbance**  The absorption of part of the visible spectrum by an object or solution. The quantitative relationship of absorbance as a function of the length of the light path and the concentration of the absorbing species is the principle behind spectrophotometry.

**agar**  An extract of red algae (family Rhodophyceae) used as a solidifying agent for microbiological media.

**alkaline phosphatase**  An enzyme that can be conjugated to biotin to function as part of the detection system for biotinylated probes. Reaction of enzyme with the substrate BCIP (5-bromo-4-chloro-3-indolyl phosphate) generates a colored product.

**ampicillin**  A semisynthetic penicillin that inhibits cell wall synthesis in bacteria. It is commonly used in molecular biology for selection of ampicillin-resistant microorganisms.

**anode**  The positive electrode, usually colored red in a gel electrophoresis apparatus. Negatively charged nucleic acid molecules migrate to the anode from the cathode when an electrical field is applied.

**aseptic technique**  A set of standard, commonsense rules of microbiological practice that ensures that procedures are executed with a minimum of risk to the worker and without contamination of the samples and work space.

**aspirate**  To remove a liquid layer such as a supernatant from a sample, using a pipet or equivalent attached to a vacuum source with a trap to catch effluent.

**biotinylation of nucleic acids**  A nonradioactive method of labeling nucleic acid probes using nick translation to incorporate biotin-derivatized nucleotides (dUTP). An alternative to radioactive labeling.

**bromphenol blue**  A chemical dye used as a pH indicator that progressively turns from yellow to blue over the range of pH 3–5. Also commonly used as a tracking dye for nucleic acids in electrophoresis gels. The dye migrates to the position of ~200 base pairs.

**bromthymol blue**  A chemical dye used as a pH indicator that progressively turns from yellow to blue over the range of pH 6–7.5.

**buffer**    A conjugate acid–base pair that functions as a system for resisting changes in pH; especially important for maintaining pH during studies of biological systems *in vitro*.

**cathode**    The negative electrode, usually colored black in a gel electrophoresis apparatus. Negatively charged DNA molecules will migrate from the cathode to the anode when the electrical field is applied.

**cesium chloride (CsCl)**    A heavy inorganic salt used for creating density gradients in a high-speed ultracentrifuge to purify nucleic acids.

**chloramphenicol**    A broad-spectrum antibiotic from *Streptomyces venezuelae* that interferes with peptide bond formation in prokaryotes. Used as a selection agent and for amplifying relaxed plasmids.

**clone**    (1) A genetically identical population of cells that has descended from one cell. (2) To purify a genetically identical population of cells from a large number of genetically heterogeneous cells. May also refer to a gene or piece of DNA.

**colony**    A visible growth of microorganisms or cells on a solid microbiological medium. A colony may or may not be a clone.

**competent**    A particular condition of cells such as *Escherichia coli* following chemical treatment to make the cell envelope permeable to exogenous DNA molecules.

**complex medium**    Nutritional medium for microorganisms that includes ingredients such as yeast extract and Bacto Peptone, for which the exact chemical composition is not known.

**cuvette**    A small plastic or quartz vessel of specific dimensions and light-absorbing qualities designed to hold a sample for spectrophotometry.

**defined medium**    A nutrient medium for microorganisms for which the exact chemical composition is known.

**denaturation of nucleic acids**    Strand separation of double-stranded nucleic acid caused by heating or treating with alkali to enable hybridization of the resulting single-stranded species (such as a labeled probe) to another single-stranded nucleic acid target.

**deoxynucleotide triphosphates (dNTPs)**    Precursor molecules used in the enzymatic synthesis of DNA.

**dialyze**    To remove impurities such as salts from a solution of macromolecules by allowing diffusion of smaller molecules across a semipermeable membrane into water or an appropriate buffer.

**DNA polymerase I**    The *E. coli* DNA polymerase enzyme that contains an excision repair function that is taken advantage of in nick translation reactions to incorporate labeled nucleotides into a DNA probe.

**electrophoresis**    A method used to separate charged molecules that migrate in response to the application of an electrical field. In gel electrophoresis, heterogeneously sized molecules in a sample are drawn through an inert matrix such as agarose and separate during migration according to size.

**3′ end**   The end of a piece of DNA (or RNA) that contains the phosphate group attached to the 3′ carbon of the corresponding base of the nucleotide.

**5′ end**   The end of the piece of DNA (or RNA) that contains the hydroxyl group attached to the 5′ carbon of the corresponding base of the nucleotide.

**ethidium bromide (EtBr)**   An orange dye (and strong mutagen) that intercalates between the stacked bases of nucleic acids (double-stranded most efficiently) and causes the molecules to fluoresce under ultraviolet light. Fluorescence allows visualization of fragments in gels and cesium chloride density gradients and also provides a way to quantitate nucleic acids whereby the intensity of fluorescence is compared to known standards.

**filter sterilize**   To remove contaminating microorganisms from liquids by suction through a particle-retaining membrane into a sterile container.

**β-galactosidase**   An enzyme encoded by the *lacZ* gene of *E. coli* that catalyzes the hydrolysis of lactose to glucose and galactose. This enzyme is used in many different molecular biology applications.

**hockey stick**   (slang) A piece of solid glass rod about 15 cm long and 5 mm in diameter that has been bent into the shape of a hockey stick. It is used to spread small volumes of solutions or cells on microbiological plates after flame sterilizing the bent end with ethanol.

**host**   The organism that has been infected or transformed with a virus or plasmid.

**hybridize**   To allow complementary strands of nucleic acid to anneal to yield a double-stranded product. Hybridization can be controlled by salt concentration and temperature (see **stringency**). Also see **Northern blot** and **Southern blot.**

**kilobase (kb)**   A unit of length for nucleic acid strands equal to 1000 bases or nucleotides.

**Klett**   Name of a colorimeter model that uses colored filters and light scattering to measure the density of a cell culture in arbitrary "Klett units."

**β-lactamase**   The enzyme, encoded by the *bla* gene, that cleaves the β-lactam–thiazolidine ring of penicillin antibiotics. Organisms containing this gene are ampicillin resistant. *bla* is commonly included on plasmid constructions as a selectable marker (usually designated *Amp^r*) for transformed hosts such as *E. coli.*

**λ *Hind*III**   A set of DNA fragments (0.5–23.1 kb) generated by digesting the phage λ with the restriction enzyme *Hind*III. The fragments are used as standard size markers on agarose gels and are commercially available.

**LB (Luria–Bertani)**   A complex rich medium for culturing bacteria.

**log phase**  Stage of exponential microbial growth, following the lag phase, when cells are dividing at a constant rate.

**loop**  A utensil used in microbiology consisting of a straight handle with a wire approximately 4 inches long. The wire can be flame sterilized and the end is twisted into a loop which is convenient for inoculating cultures, streaking plates, etc.

**lysozyme**  An enzyme that removes the bacterial cell wall and causes cell lysis in the presence of EDTA. Used in plasmid DNA isolation and purification procedures.

**map**  (1) To determine the linear order of restriction sites on a piece of DNA by performing a series of digests and analyzing the fragment banding pattern on a gel. (2) The order of the restriction sites generated by mapping.

**master plate**  A plate that contains the original microbial colonies from which replica plates were made.

**maxi-prep**  A large-scale procedure for the isolation and purification of plasmid DNA for 100-ml to 1-liter cultures of *E. coli* yielding milligram quantities of pure plasmid.

**microcentrifuge**  A small tabletop centrifuge with a radius of 40–50 mm and rotor holes for microcentrifuge tubes that is capable of speeds up to 14,000 rpm.

**mini-prep**  A small-scale procedure for the isolation and purification of plasmid

DNA from an *E. coli* host, which entails growth and lysis of the bacteria, differential centrifugation in the microcentrifuge, and purification.

**nick**  A small gap in one strand of double-stranded DNA caused by mechanical stress, UV light, or an enzyme such as DNase I.

**nick translation**  An *in vitro* method for labeling DNA. *Escherichia coli* DNA polymerase I can use a nick as a starting point to excise nucleotides progressively and to replace them with labeled nucleotides. The nick progresses or is "translated" as the polymerase moves 5′ to 3′ along the DNA template until another nick is encountered. ("Translation," as used here, has nothing to do with RNA translation into protein.)

**nitrocellulose**  A transfer medium on which nucleic acids or proteins are immobilized by blotting.

**Northern blot**  (1) To hybridize DNA to RNA that was transferred electrophoretically or by capillary action from a gel onto an appropriate medium such as nitrocellulose or nylon membrane. (2) The piece of transfer medium that now contains RNA sequences that have been blotted.

**nylon membrane**  A transfer medium for blotting nucleic acids or protein that is stronger and more versatile than nitrocellulose.

**oligonucleotide primer**  A chemically synthesized chain of ~10–50 nucleotide bases typically used for DNA sequencing or amplification by PCR.

**plasmid** A self-replicating circle of extra-chromosomal DNA. Plasmids are used in molecular biology as cloning vectors to introduce foreign DNA into a host cell. Plasmids occur naturally in bacteria and usually impart some selective advantage to the host such as antibiotic resistance.

**plate** (1) A petri dish (2) To spread cells onto a solid nutrient medium in a petri dish.

**polyethylene glycol (PEG)** An organic reagent used to alter the cell envelope of yeast to facilitate transformation.

**polymerase chain reaction (PCR)** A method for the amplification of specific regions of DNA using oligonucleotide primers, dNTPs, and a heat-stable DNA polymerase.

**probe** (1) A piece of labeled DNA or RNA used to locate immobilized sequences on a blot by hybridizing under optimal conditions of salt and temperature. (2) To hybridize a probe to a blot.

**quadrant streak** A style of streaking for isolated single colonies of microorganisms on solid medium that entails dividing the plate into four areas or quadrants.

**restrict** To digest with a restriction enzyme.

**restriction enzyme** An enzyme that cleaves double-stranded DNA at specific recognition sequences of nucleotides; used extensively in standard molecular biology procedures such as mapping or modifying DNA for cloning.

**restriction site** The recognition sequence for a restriction enzyme. May specifically refer to the exact point of cleavage between nucleotides within the sequence.

**selectable marker** A gene that permits survival of a selected phenotype. Plasmids usually contain selectable markers such as antibiotic resistance genes for identification of transformants from an antibiotic sensitive background of cells cultured on selective medium.

**Southern blot** (1) To hybridize DNA to DNA that was transferred electrophoretically or by capillary action from a gel onto an appropriate medium such as nitrocellulose or a nylon membrane. (2) The piece of transfer medium that now contains the DNA sequences that have been blotted.

**sterilization** The process of eradicating all life forms. Culture medium, glassware, and utensils are sterilized before use to prevent contamination. Accomplished by autoclaving, irradiating, heating, etc.

**stop buffer** A buffer containing a component such as the metal-chelating agent EDTA that will stop an enzymatic reaction. Added to restriction digests at the end of incubation or sometimes included in the tracking dye solution.

**streak** To use a loop, toothpick, or other sterile utensil to dilute out microorganisms on solid medium in an effort to obtain isolated colonies.

**streptavidin**   A biotin-binding protein that is a component of the detection system for biotinylated probes; it is capable of binding to both the probe (containing biotinylated dATP) and the biotinylated alkaline phosphatase because it is tetravalent.

**stringency**   Conditions used in the hybridization of nucleic acids that reflect the degree of complementarity between the hybridizing strands. Parameters of high stringency are low salt and high temperature, wherein highly complementary strands hybridize; low stringency conditions of high salt and low temperature increase overall hybridization.

**subculture**   To reinoculate fresh culture medium with cells from an existing culture, such as from an overnight culture or streak plate.

**TE [10 m$M$ Tris–1 m$M$ EDTA (pH 7.5–8.0)]**   A buffer used in molecular biology for operations, such as dialyzing or solubilizing DNA, where only a dilute buffer is required.

**$T_m$**   The melting temperature of a nucleic acid hybrid. The $T_m$ is related to the G + C content, and also to the temperature and ionic strength of the washing buffer.

**tracking dye**   A solution that is mixed with samples to be loaded onto a gel. It contains an agent, such as glycerol or Ficoll to sink the sample into the well, and dyes, such as bromphenol blue, that migrate at a known rate so that the migration of electrophoresed samples can be followed.

**transformation**   The process by which microorganisms can accept and incorporate exogenously added DNA. Transformation occurs naturally among microbes or can be artificially induced using competent cells.

**Tris   [tris(hydroxymethyl)aminomethane]**   An organic buffer used in molecular biology, biochemistry, and cell biology.

**Triton X-100**   A nonionic detergent.

**unit of activity**   Generally, the amount of an enzyme that catalyzes the conversion of a certain quantity of substrate to product in a fixed time period. For a restriction enzyme, 1 unit is defined as the amount required to digest 1 $\mu$g of standard DNA, such as phage $\lambda$, in 1 hour at optimal temperature and buffer conditions in a specified volume.

***URA3***   A gene used as a selectable marker for yeast.

**Western blot**   (1) To use antibodies to detect immobilized proteins. (2) The piece of material onto which the protein was immobilized and probed with antibodies.

**YEPD (yeast extract–peptone–dextrose)**   A complex medium for culturing yeast.

**YNB (yeast nitrogen base)**   A minimal medium for culturing yeast cells.

Absorbance, 174–176, 247; *see also*
    Optical density
Acids, properties of, 200
Acrylamide-bis
    preparation of, 151
    safety precaution, 150
Adenine sulfate, as medium
    component, 22
Addresses, manufacturers and
    distributors, 237–244
Aeration, importance of, for cultures, 215
Agar, as solidifying agents in medium, 18
Agarose gel electrophoresis, 63–74
    buffer for, 66
    for determination of purity and size of
        DNA, 44–45
    theory of, 63–65
Agarose gels
    minigels, 65
    photography of, 71–72
    Southern blotting from, 75–84
    staining of, 71–73
Alkaline phosphatase, 86; *see also* Biotin-
    modified alkaline phosphatase
Amino acids
    pools in minimal medium, as test for
        auxotrophy, 166
    preparation of stock solutions,
        212–213
Ammonium acetate, for DNA
    precipitation, 40
Ammonium persulfate, in SDS–PAGE
    gels, 151, 153

Ammonium sulfate, for precipitation for
    β-galactosidase, 46–47
Ampicillin
    maintenance of selective pressure
        with, 32
    preparation of stock solution, 32
    for selection of transformants, 112
Amplification, of plasmid DNA, 49–54,
    178–188
Antibiotics
    ampicillin
        maintenance of selective pressure
            with, 32
        preparation of, 32
        for selection of transformants, 112
    chloramphenicol
        amplification of plasmids with, 178
        preparation of, 179
Aseptic technique, 9–11
Autoclave, use of, 20–21
Auxotrophic mutants, isolation and
    characterization of, 165–169

*Bam*HI, 66, 68, 73
Bases, properties of, 200
BCIP solution, 89, 94, 247
Biotin-modified alkaline phosphatase, 86
Biotinylated DNA probes
    preparation of, 88
    purification of, from unincorporated
        nucleotides, 89

Blue Dextran, 144–146
Bovine serum albumin
    in protein assay, as standard,
        121–123
    in restriction digests, 65–68
Bradford dye, preparation of, 123
Bradford protein assay, 120, 122–123
BRL DNA detection system, components
        of, 88
BRL Nick Translation System,
        components of, 88
Bromphenol Blue
    relationship between optical density and
        concentration, 175–177
    SDS-PAGE sample buffer, component
        of, 151
    tracking dye, component of, 66, 247
Bromthymol blue, as pH indicator,
        171–172, 247
Buffers
    for agarose gel electrophoresis, 66
    for blotting DNA, 77–78
    hybridization, 88
    for mini-preps, 32
    phosphate, 171–172
    prehybridization, 89
    for preparation of probe, 88
    restriction enzyme
        reaction, 66
        stop, 66
    for SDS–PAGE, 151
    for Southern transfer, 77
    TE, 33
    theory of, 192–194, 195–198
    for transformation
        Escherichia coli, 105–112
        Saccharomyces cerevisiae, 97–103
    Tris, 228–229
    Z buffer, 127

Calcium chloride
    for transformation of Escherichia coli,
        107–108
Cell extraction methods, 136
Cell mass, determination of, 219–220
Cell number, determination of,
        221–222
Cell viability, determination of, 217–218
Cesium chloride
    density-gradient centrifugation with,
        178–188
    safety precaution, 183–184
Chloramphenicol
    for ampliciation of plasmid DNA,
        178
    preparation of stock solution, 179
Chloroform, extraction of solutions of
        DNA and RNA with, 38–40
Colony hybridization, 189–191
Column chromatography; see Solid-phase-
        anion-exchange columns, Gel
        filtration chromatography
Competent cells
    Escherichia coli, 107–108
    Saccharomyces cerevisiae, 97–102
Complex medium
    definition of, 17–20, 248
    LB, preparation of, 33
    YEPD, preparation of, 19–21
Concentration of DNA, by precipitation
        with
    ethanol, 37–41
    isopropanol, 37–41
Continous streak, 14
Coomassie Brilliant Blue
    Bradford dye, component of, 120
    stain for SDS–PAGE, component of
        staining solution, 151–152
Culture transfer, 14–15

Death phase, of microbial growth, 25, 214

Defined medium
  definition of, 17
  preparation of components for, 212–213
  YNB, for *Saccharomyces cerevisiae*, 21–22

Denaturation buffer, composition of 5× stock, for Southern transfer, 77

Denhardt's solution, preparation of, for Southern transfer, 89

Destaining of SDS gels, 151

Dialysis tubing
  general preparation of, 180–181
  for removal of ammonium sulfate, 149
  for removal of cesium chloride from DNA solutions, 186

Dichlorodimethylsilane, to coat glassware, 226

Diethyl pyrocarbonate, for removal of RNases from glassware, 226

Dilutions, proper execution of, 201–202

Dimethyl sulfoxide, 208

Dithiothreitol, 65, 67

DNA polymerase I of *Escherichia coli*, nick translation with, 87

DNA, storage of, 208–209

DTT, *see* Dithiothreitol

*Eco*RI, 66, 68, 73

EDTA, preparation of stock solution, 180–181, 229–230

Electrophoresis, 63–74; *see also* Agarose gel electrophoresis for SDS–PAGE

Enzymes; *see also* β-galactosidase
  alkaline phosphatase, 86
  *Bam*HI, 66, 68, 73

*Eco*RI, 66, 68, 73
*Hind*III, 66, 68, 70, 73
  lysozyme
    for mini-preps, 32
  *Pst*I, 66, 68, 73
  rules for handling, 231–234

*Escherichia coli*, 26, 32–33, 105–112, 207–209, 219, 223–225
  growth medium for, 33
  nomenclature, 223–225
  relationship of optical density to cell mass, 219
  sources of, 207
  storage of, 208–209
  strain LE392, 32, 207
    doubling time, 26
    genotype, 32
    transformation of, 105–112
  strain LE392(pRY121), 207
  colony hybridization with, 189–191
  maxi-prep of DNA from, 49–54
  mini-prep of DNA from, 31–36

Ethanol, for precipitation of DNA, 38–41

Ethidium bromide
  in cesium chloride density gradients, 180, 184–187
  inactivation of, 42–43
  intensity of fluorescence, to quantitate DNA, 43–44
  preparation of, 42
  safety precaution, 41
  as stain for agarose gels, 71–73

Ethylenediaminetetraacetic acid, *see* EDTA

Exponential phase, of microbial growth, 25, 214–215

Folin-Ciocalteu's reagent, 121

β-Galactosidase
  assays for
    in cell extracts, 135–138
    in permeabilized yeast cells,
      131–134
    in yeast colonies, 125–129
  molecular weight of, 150
  purification of, 141–156
    by ammonium sulfate precipitation,
      146–149
    by gel filtration chromatography,
      146–149
  specific activity of, 138
  units of activity, calculation of,
    138
GAL promoter, 72
Gel filtration chromatography
  large-scale preparation of plasmid
    DNA, 49–54
  Sephadex G-50
    for recovery of labeled DNA
      probe, 91
  Sephadex G-200
    calibration of column, 144–146
    purification of β-galactosidase,
      149
  theory of, 142–144
Gels; see Agarose gels, SDS–PAGE gels
Glass beads, for breakage of yeast cells,
  135–139
Glycerol
  as cryogenic agent, 182, 208–209
  effect on restriction enzyme activity,
    67
  SDS-PAGE sample buffer, component
    of, 151
Growth aeration, 215
Growth curve, 25–30, 214–216
Growth rate constant, 215

Hemocytometer
  for cell enumeration, 26, 28, 136–137
  instructions for proper use, 221–222
Henderson–Hasselbalch equation, 193
HindIII, 66, 68, 70, 73
Histidine
  as medium component, 22
  preparation of, 213
Hybridization, of probe, to Southern blot,
  85–96, 249

Inoculation, of cultures, 9–16, 215–216
Isoamyl alcohol
  for extraction of ethidium bromide
    from DNA solutions, 180, 186
  mixed with chloroform, for extraction
    of protein from DNA solutions,
    38, 40
Isolation, of single colonies, on agar
  by continous streak, 14
  by quadrant streak, 13–14
Isopropanol, for precipitation of DNA,
  32, 34

Klett colorimeter, 176, 249

λ phage
  digested with HindIII, as molecular size
    markers in agarose gels, 70, 72,
    77, 249
  as quantitative DNA standards, 41,
    45–46

Labeling DNA, by nick translation, 85–93

$\beta$-Lactamase, 32, 249

*lacZ* gene, 125

Lag phase, of microbial growth, 25, 214

LB medium, preparation of, 33

LE392, strain of *Escherichia coli*, genotype of, 32

Lithium acetate/TE, 99, 101

Lithium chloride/TE solution, composition of, 99

Log phase, of microbial growth, 25, 214, 250

Lowry protein assay, 119–123, 235–236

Luria-Bertani medium, *see* LB medium

Lysine
    as medium component, 22
    preparation of, 212

Lysis buffer, for mini-preps preparation of, 32

Lysozyme, preparation of working solution
    for mini-preps, 32

Magnesium chloride, for improvement of DNA precipitation, 41

Maxi-prep, of DNA, 178–188

Media, microbiological, preparation of
    for *Escherichia coli*
        LB, 33
    for *Saccharomyces cerevisiae*
        YEPD, 19–21
        YNB, 21–22

2–Mercaptoethanol, 150–151

N-Methyl-N'-nitro-N-nitrosoguanidine, *see* MNNG

Microorganisms; 201–222, *see also* *Escherichia coli, Saccharomyces cerevisiae*
    determination of
        cell mass, 219–220
        cell number, 221–222
        viability, 217–218
    inoculation and subculture, 215–216
    list of, for this course, 207
    nomenclature, 223–225
    safe handling, 204–206
    storage of, 208–209

Micropipettor, proper use of, 201–202

Minigels, 63–74

Minimal media
    YNB, for *Saccharomyces cerevisiae*, 21–22

Mini-prep
    alkaline lysis method, 113–116
    boiling method, 31–36

MNNG
    preparation of, 167
    yeast mutagenesis with, 165–166

Mutagens, types of, 165

NBT solution, 89

Neutralization buffer, composition of, for Southern transfer, 77

Nick translation, 85–93

Nitrocellulose
    colony hybridization, 189–191
    fixation of DNA to, 83
    Southern blotting, transfer medium for, 75, 81, 83–84

*o*-Nitrophenol, function in $\beta$-galactosidase assay, 133

o-Nitrophenol-β-D-galatopyranoside, *see* ONPG
Nonmenclature, for microorganims, 223–225
Nylon membranes,
    colony hybridization, 189–191
    fixation of DNA to, 83
    Southern blotting, transfer medium for, 75, 81, 83–84

OD, *see* Optical density
ONPG, substrate for β-galactosidase, 132–133, 136
Optical density
    for measurement of cell mass, 26–30, 52, 108, 132, 137, 182, 219–220,
    for measurement of β-galactosidase, 133, 138
    for quantitation of DNA, 38, 43
    for quantitation of protein, 122–123, 146

PAGE, *see* SDS–PAGE gels
Peptone, component of YEPD, 19
pH
    definition of, 192–194
    measurement of, 170–173
Phenol
    effect on plasticware, 227
    extraction of protein from DNA solutions with, 35–40
    preparation of, 39
    safety precautions, 39
Phenol-chloroform, 40
Phenylmethylsulfonyl fluoride, *see* PMSF
pH meter, 170–173

Phosphate buffers, 170–173, 192, 194
Photography, of agarose gels, 42–46
Plasmid DNA
    chromatographic purification of, 49–54
    concentration of, 40–41
    maxi-prep of, 178–188
    mini-prep of, 31–36
    precipitation of, 33–35
    quantitation of, 41–46
    RNA contamination of, 63–65
    solvent extraction of, 38–40
    transformation with
        *Escherichia coli*, 105–112
        *Saccharomyces cerevisiae*, 97–103
Plasmids
    amplification of, 49–54, 178–188
    definition of, 251
    pRY121, map of, 72
    restriction digestion of, 63–74
PMSF, preparation of, 136
Polyacrylamide gel electrophoresis, *see* SDS–PAGE gels
Polyethylene glycol, for transformation of yeast, 99
Polymerase chain reaction (PCR), 37, 55–62, 111
Potassium phosphate, dibasic and monobasic, part of inorganic buffer system, 171–173
Precipitation of DNA, *see* Concentration of DNA
Prehybridization buffer, preparation of 1.5× stock, 89
Probes, DNA
    biotinylated
        preparation of DNA, 90–91
        purification of DNA, 90–92
        verification of, 93–95
    hybridization to Southern blot, 95–96

Protein, assays for
  Bradford, 120, 122–123
  in crude cell extracts, 137
  in large-scale chromatographic column
      eluent, 149
  Lowry, 119–123, 235–236
pRY121; *see also* Plasmid DNA,
      Plasmids
  restriction map of, 72
  as shuttle vector, 32, 99
*Pst*I, 66, 68, 73
Pure culture
  definition of, 9
  establishment of, 9–16
Purines
  medium component, as test for
      auxotrophy, 166
  preparation of, 213
Pyrimidines
  medium component, as test for
      auxotrophy, 166
  preparation of, 213

Quadrant streak, 13–14
Quantitation of DNA, 37–46
  by agarose gel method, 44–46
  by ethidium bromide fluorescence,
      43–44
  by UV absorption, 43
Quick-Seal tubes, 180, 184

Records, laboratory, 3, 5
Replica plating, 166–169
Restriction digestion, of DNA,
      63–74

Restriction enzymes, *see also Bam*HI,
      *Eco*RI, *Hin*dIII, *Pst*I
  digestion of DNA with, 63–74
  reaction buffers, composition of, 66
RNase
  preparation of, 33
  removal of RNA from DNA
      preparations with, 35, 115
Running buffer, for SDS–PAGE,
      composition of, 151

*Saccharomyces cerevisiae*
  growth curve, 25–30
  growth in liquid medium, 29–30,
      214–216
  nomenclature, 224–225
  relationship of optical density to cell
      mass, 219
  sources of, 207
  storage of, 208–209
  strain X2180–1A, 207
      doubling time for, 26
      genotype of, 26
      transformation of, 97–103
  strain YNN281 transformed with
      pRY121, 207
      colony hybridization of, 189–191
      extraction of, 109–112
      permeabilization of, 131–134
Safety precautions
  cesium chloride, 183–184
  ethidium bromide, 41
  glass wool, 91
  handling of microorganisms, 204–206
  laboratory rules for, 7
  phenol, 39
  ultraviolet lights, 44

Salmon sperm DNA
  denatured, as filter blocking agent, 89
Salmon testes DNA
  restriction enzyme digestion of, 77
Sample buffer, for SDS–PAGE,
    preparation of 2× stock, 151
Sarkosyl, for permeabilization of cells,
  131
SDS–PAGE gels, 149–150
  destaining, 151
  electrophoresis of, 154
  fixing, 154–155
  preparation and pouring, 150–153
  staining of, 154–155
Selection of transformants
  of *Escherichia coli*, with ampicillin, 112
  of *Saccharomyces cerevisiae,* by uracil
    auxotrophy, 102
Selective pressure, maintenance of
  by ampicillin, 32
  by growth in medium without uracil,
    102
Separating gel buffer, for SDS–PAGE
  composition of, 151
Sephadex G-50
  for recovery of labeled probe, 87, 91
Sephadex G-200
  preparation of, 144
  for purification of β-galactosidase,
    142–146
Sf solution, composition of, 132
Shuttle vector, 98
Sodium acetate, for precipitation of
  DNA, 33
Sodium azide, as antimicrobial agent, 145
Sodium bicarbonate, for preparation of
  dialysis tubing, 180
Sodium carbonate to stop β-galactosidase,
  132–133, 136, 138

Sodium chloride
  neutralization buffer, component of, 77
  SSC, component of, 78
Sodium dodecyl sulfate (SDS), preparation
    of stock solution, 95
Sodium lauroyl sulfate (Sarkosyl), for
    permeabilization of cells, 131
Sodium thiosulfate, 168
Solid-phase anion-exchange columns,
  50–54
  for large-scale preparation of DNA,
    50–54
Solutions, preparation of, 195–199
Southern blotting, 75–84
Specific activity
  definition of, for an enzyme, 141
  β-galactosidase, 138
Spectrophotometry
  applications of, 174–175
  for determination of DNA purity, 43
  for measurement of
    cell mass, 26–30, 52, 108, 132, 137,
      182, 219–220
    column eluent, 146
    o-nitrophenol, 133, 138
    protein concentration, 122–123, 146
  for quantitation of DNA, 38, 43
  theory of, 174–177
SSC, composition of 20X stock, 78
Stacking gel buffer, composition of, for
  SDS-PAGE, 151
Stationary phase, of microbial growth,
  25, 214
Sterile technique, *see* Aseptic technique
Sterilization
  by autoclaving, proper use of, 20–21
  methods of, 210–211
Stop buffer, composition of, for
    restriction digests, 66, 251

Streak plate, 9–15
Streptavidin, 85–86, 89, 252
Stringency, 57
Sucrose, as component of tracking dye, 66
Synthetic medium, *see* Defined medium

$T_m$, 87, 252
*Taq* polymerase, 56–58
TEA buffer, composition of, 66
TE buffer, composition of, 33
TEMED, 152–153
Toluene, for permeabilization of microbial
    cells, 131–133
Tracking dye, composition of 6× stock,
    for agarose gels, 66
Transfer media, for Southern blotting, 75
Transformation
    definition of, 252
    efficiency of, 98, 116
    of *Escherichia coli*, 105–112
    of *Saccharomyces cerevisiae*, 97–103
Tris buffers, preparation of stock
    solutions, 228–229
Tris-sucrose solution, for maxi-prep,
    composition of, 180
Triton X-100 solution
    for maxi-prep, composition of, 180
    for preparation of probe, 88

Ultraviolet light
    damage to DNA, 72
    for fixation of DNA to nylon
        membranes, 83
    for location of plasmid band in cesium
        chloride density gradients, 185

for nicking DNA to facilitate Southern
    transfer, 75, 80
for quantitation of nucleic acids bound
    with ethidium bromide, 43–44
safety note, 44
for visualization of DNA fragments in
    agarose gels, 44–46
*URA3*, 72, 252
Uracil
    auxotrophy, for selection of
        transformants, 102
    as medium component, 22

Washes, of Southern blot,
    posthybridization to probe, 95
Western blot, 157–161, 252
Whole cell DNA, for Southern
    transfer, 77

X-gal, 5-bromo-4-chloro-3-indolyl-$\beta$-D-
    galactopyranoside, 126

Yeast extract
    component of LB, 33
    component of YEPD, 17–19
Yeast extract-peptone-dextrose, *see* YEPD
Yeast nitrogen base, *see* YNB
YEPD medium, preparation of, 19–21
YNB medium, preparation of, 21–22

Z buffer, composition of, 127